AF587412

Back to the Moon

Navigating the Artemis Journey

Space Science, Exploration and Policies Series

The National Aeronautics and Space Administration's Role in Space Exploration and Development
Brent M. Jenkins (Editor)
ISBN: 979-8-89530-539-3
eBook ISBN: 979-8-89530-564-5

Commercial Human Spaceflight: Background, Regulations and Safety
Dana G. Pritchard (Editor)
ISBN: 979-8-89530-176-0
eBook ISBN: 979-8-89530-195-1

A Guide to Black Holes
Kenath Arun, PhD
ISBN: 979-8-88697-163-7
eBook ISBN: 979-8-88697-515-4

Conceptual Features of Einstein's Theory of General Relativity Based on the Philosophy of Science
Jun-Young Oh
ISBN: 979-8-88697-107-1
eBook ISBN: 979-8-88697-140-8

Time-Out of Time: A Postscript to Nuclear Time Travel
Bernd Schmeikal, PhD (Author)
ISBN: 978-1-68507-320-6
eBook ISBN: 978-1-68507-465-4

More information about this series can be found at
https://novapublishers.com/product-category/series/space-science-exploration-and-policies/

Silas Blackwood
Editor

Back to the Moon

Navigating the Artemis Journey

Library of Congress Cataloging-in-Publication Data

ISBN: 979-8-89530-861-5 (Hardcover)
ISBN: 979-8-89530-962-9 (eBook)

Published by Nova Science Publishers, Inc. † New York

Contents

Preface

For decades, Congress has prioritized adolescent pregnancy prevention to address the public health, social, and economic challenges it poses to adolescents, their children, and society as a whole. Since the 1980s, Congress has authorized programs—administered by the U.S. Department of Health and Human Services (HHS)—designed to reduce adolescent pregnancy rates. Chapters 1 and 2 explore the evolution and status of four federally funded adolescent pregnancy prevention programs.

Adolescent childbearing carries significant social, health, and financial risks, not only for young parents but also for their families and broader communities. Despite steady declines in teen birth rates, as reported by the National Center for Health Statistics (NCHS) within the Centers for Disease Control and Prevention (CDC), the United States continues to report some of the highest rates among industrialized nations. Chapter 3 examines these teen birth trends.

Young people in the child welfare system face unique challenges, which are magnified for those who are pregnant or parenting. Chapters 4 and 5 provide strategies for caseworkers, agencies, and youth-serving professionals to better support these individuals and their families.

Federal policies also play a pivotal role in ensuring the rights and protections of pregnant individuals. Recent updates to Title IX regulations, as covered in Chapter 6, aim to eliminate discrimination against pregnant students and employees in educational institutions. Chapter 7 details the Pregnant Student Rights Act, which mandates that higher education institutions provide pregnant students with information about their rights, accommodations, and resources.

In the workplace, various federal laws safeguard employees from pregnancy-related discrimination. Chapters 8 through 11 offer an overview of key protections, such as the Pregnancy Discrimination Act (PDA), while discussing their legislative history, practical applications, and limitations. Lastly, Chapters 12 and 13 focus on resources for pregnant individuals with substance use disorders, providing valuable information for both affected individuals and the professionals who support them.

Chapter 1

Artemis: NASA's Program to Return Humans to the Moon*

Rachel Lindbergh

Between 1969 and 1972, the Apollo program of the National Aeronautics and Space Administration (NASA) landed 12 American men on the Moon and returned them safely to Earth (see Figure 1). Since then, no human has been farther from Earth than low-Earth orbit, a few hundred miles up; the distance to the Moon is about 240,000 miles. Artemis, named for Apollo's twin sister in ancient Greek mythology, is NASA's program for a return to the Moon by American astronauts—one of them a woman—in 2026.

Orion and the Space Launch System

Artemis has evolved from plans initiated in the NASA Authorization Act of 2010 (P.L. 111-267). The act established a statutory goal of "expand[ing] permanent human presence beyond low-Earth orbit" and mandated the development of a crew capsule and a heavy-lift rocket to accomplish that goal. The capsule, now known as Orion, and the rocket, known as the Space Launch System (SLS), have been in development since then (see Figure 2).

---

* This is an edited, reformatted and augmented version of Congressional Research Service Publication No. IF11643, updated February 8, 2024.

In: Back to the Moon
Editor: Silas Blackwood
ISBN: 979-8-89530-861-5

Source: NASA, https://www.nasa.gov/image-detail/amf-as17-134- 20382/.
Note: This image shows Apollo 17 astronaut Harrison Schmitt standing on the surface of the Moon on December 13, 1972. Behind him are the Lunar Module lander and the Lunar Roving Vehicle rover.

Figure 1. The Last Human Lunar Mission: Apollo 17.

Each Orion capsule consists of a crew module with room for four to six astronauts as well as storage space and a docking port; a service module (contributed by the European Space Agency) to provide power and propulsion; and a launch abort system. The crew module is the only portion intended to return to Earth at the end of a mission; it is designed to be reusable.

The SLS is an expendable rocket designed to carry Orion into space and set it on its initial trajectory. The SLS could also potentially be used for other missions involving heavy payloads or requiring very high thrust. It is designed to be upgraded in stages (known as Block 1, Block 1B, and Block 2) by substituting improved versions of its major elements. For example, for Block 1B, NASA is developing the Exploration Upper Stage to replace the Block 1 upper stage, which is known as the Interim Cryogenic Propulsion Stage.

In December 2014, a partially complete Orion was launched on a Delta IV Heavy rocket and orbited Earth twice before splashing down in the Pacific Ocean. This uncrewed mission tested the crew module's heat shield and parachutes, as well as other systems.

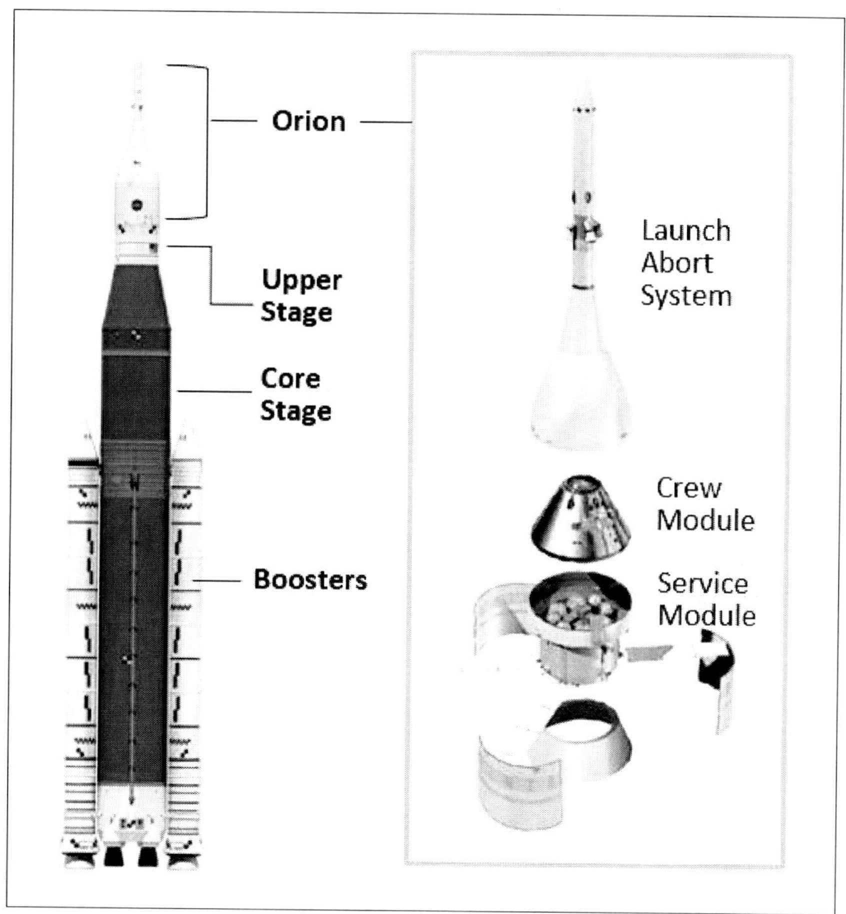

Source: CRS illustration based on NASA diagrams at https://www.nasa.gov/directorates/esdmd/space-launch-system-ftdku/ and https://oig.nasa.gov/docs/IG-20-018.pdf.

Figure 2. Major Elements of SLS and Orion.

The first launch of Orion on an SLS was in November 2022. During this mission, known as Artemis I, a complete but uncrewed Orion orbited the Moon before returning to Earth. The mission was designed to provide the data NASA needs to certify safety for crewed flights.

Artemis II, the first crewed test of Orion and the SLS, is expected in 2025. During this 10-day mission, Orion and its crew of four are to fly around the Moon at an altitude of about 4,000 miles before returning to Earth.

The Artemis III mission, planned for 2026, is to include the first human Moon landing since 1972. Achieving that goal would require the development of other systems, such as a lunar lander.

Subsequent Artemis missions, with more lunar landings and various additional capabilities, are planned approximately every year starting in 2028.

Human Landing System

The Orion capsule is not designed to land on the Moon. Instead, for Artemis III and subsequent lunar surface missions, astronauts will need to transfer to a separate spacecraft, known as a Human Landing System (HLS), for lunar descent and ascent. In April 2021, NASA selected SpaceX to provide an HLS as a commercial service starting with Artemis III. In May 2023, it awarded a second HLS contract to Blue Origin to provide an alternative to the SpaceX system starting with Artemis V (planned for 2029). Both systems are still in development.

Gateway

To facilitate Artemis lunar landings and other missions, NASA is developing a modular platform, known as Gateway, to be placed in a permanent orbit around the Moon. The first two Gateway modules—the Power and Propulsion Element (PPE) and the Habitation and Logistics Outpost (HALO, a pressurized habitat for astronauts)—are currently in development, with launch planned in 2026.

Gateway is intended to serve as a depot for storing supplies, a platform for science experiments, a location where subsystems launched separately can be assembled and integrated, and a rendezvous point where astronauts can transfer between Orion and the HLS and potentially, at some point in the future, depart for Mars. NASA initially planned for Gateway to be the Orion-HLS transfer point for the Artemis III lunar landing. It is no longer part of the plans for Artemis III, but it is to be used for subsequent missions starting with Artemis IV (planned for 2028).

Other Elements

In addition to Orion, SLS, HLS, and Gateway, NASA is planning robotic missions to demonstrate new technologies and explore potential landing sites, developing new spacesuits, and developing technologies for lunar surface power, in situ use of lunar resources such as water, and other lunar surface systems such as rovers and habitats for missions after Artemis III.

Issues for Congress

As Congress oversees the progress of the Artemis program and acts on NASA authorization and appropriations legislation, it may address issues such as the schedule for the first landing, cost concerns for the program as a whole, the relative exploration priority of the Moon versus Mars, and the role of the commercial space sector.

Schedule

As recently as early 2019, NASA was planning the first post-Apollo human lunar landing for 2028. An acceleration to 2024 was announced by Vice President Pence in March 2019. Supporters of the 2024 goal argued that it instilled a sense of urgency, focus, and motivation, and that the U.S. space program is in competition with Russia and China.

Opponents argued that the 2024 date was driven by political goals rather than by technical or scientific considerations.

Congress deliberated questions such as what geopolitical or other benefits a 2024 landing might bring; how providing the funding needed to achieve a 2024 landing might affect the availability of funding for other NASA programs; how schedule pressure might influence safety decisions; and how design choices made to meet the 2024 deadline might affect system reusability for subsequent NASA human exploration missions.

NASA's schedule for Artemis III (and subsequent Artemis missions) has since slipped several times, less due to these policy debates than to development challenges. As of January 2024, the estimated Artemis III launch date is September 2026.

Budget

According to the Government Accountability Office, as of January 2024 NASA had not yet estimated the cost of the Artemis III mission or subsequent Artemis missions.

Likewise, NASA "does not plan to measure the production costs for the SLS rockets that constitute a significant proportion of future Artemis-related costs." Absent more information on long-term costs, Congress typically makes budget decisions on an annual basis. The FY2024 request for Artemis systems is $8.0 billion.

Moon or Mars?

Is returning to the Moon the primary goal for human space exploration, or is it an interim step to gain experience for future expeditions to Mars? While this distinction is to some extent a matter of emphasis, the debate continues. For example, the NASA Authorization Act of 2020 (H.R. 5666, 116th Congress) would have stated that "the Nation's human space exploration goal should be to send humans to the surface of Mars," although "reducing the risk and demonstrating the capabilities and operations needed to support a human mission to Mars may require human exploration of the cis-lunar vicinity [i.e., the region around the Moon and between Earth and the Moon] and lunar surface." This debate may drive how Artemis missions are planned (e.g., whether lunar habitats are designed to be permanent and whether potential reuse for Mars missions is a major factor in technology choices for lunar missions).

Role of the Commercial Space Sector

In recent years, NASA has placed growing emphasis on procuring services from the commercial space industry. For example, where it used to use NASA-owned space shuttles to carry cargo and crews to the International Space Station, it now buys cargo and crew transport as a commercial service on commercially owned spacecraft.

Orion and the SLS are being developed as NASA-owned systems, but the HLS is to be provided as a commercial service, and several Artemis-related robotic missions are being conducted through the Commercial Lunar Payload Services (CLPS) initiative. Not all stakeholders support the commercial approach. For example, in January 2024, former NASA Administrator Michael D. Griffin testified that the emphasis on commercial services is "the fundamental flaw in the Artemis acquisition approach." The NASA Authorization Act of 2022 directed that Artemis human landing missions are to be "carried out solely by government astronauts" (P.L. 117-167, §10811).

Acknowledgments

Former CRS Specialist Daniel Morgan wrote the original version of this product.

Chapter 2

Returning to the Moon: Keeping Artemis on Track*

Committee on Science, Space, and Technology

Purpose

The Artemis program is the National Aeronautics and Space Administration (NASA) effort to return United States astronauts to the lunar surface. The purpose of this hearing is to monitor progress on Artemis objectives, identify and understand challenges faced by NASA, and discuss the agency's path forward. This hearing also will provide the Committee with valuable insight on how NASA plans to ensure a successful American return to the Moon and enable future exploration of Mars and beyond.

Witnesses

- Ms. Catherine Koerner, Associate Administrator, Exploration Systems Development Mission Directorate, National Aeronautics and Space Administration
- Mr. William Russell, Director, Contracting and National Security Acquisitions, U.S. Government Accountability Office

* This is an edited, reformatted and augmented version of the Committee on Science, Space, and Technology, Space and Aeronautics Subcommittee Hearing Charter "Returning to the Moon: Keeping Artemis on Track" Publication, dated January 17, 2024.

In: Back to the Moon
Editor: Silas Blackwood
ISBN: 979-8-89530-861-5

- Mr. George A. Scott, Acting Inspector General, National Aeronautics and Space Administration
- Dr. Michael D. Griffin, Co-President, LogiQ, Inc

Overarching Questions

- What elements of the Artemis program are on the critical path to launching Artemis II and III? What is the status of these elements?
- NASA recently announced the delay of Artemis II and III from 2024 and 2025 respectively, to 2025 and 2026. What was the cause of this change in schedule? What is the anticipated impact of this delay on Artemis program costs?
- How has NASA adapted to mitigate future schedule and cost overruns?
- What are the most significant technical challenges NASA has faced so far? What impact have NASA management decisions had on program execution? How is NASA learning from these experiences to better execute the Artemis program moving forward?
- Among the existing risks and challenges to the Artemis program, which present the highest risks to program cost, schedule, and mission success?

Background

The Artemis program represents the next generation of United States human space exploration beyond Earth orbit. While the immediate goal of the program is to land humans on the lunar surface for the first time since the Apollo program, the Artemis program seeks to establish sustainable, long-term access to the Moon. Doing so will both advance exciting scientific research and serve as a proving ground for future human space missions to Mars and other deep space destinations.

Today's Artemis program is the result of almost two decades of evolution that started in 2004. President George W. Bush and then-Administrator Sean O'Keefe released a Vision for Space Exploration establishing goals for the

United States space program and calling for a return to the Moon.[1] Congress incorporated the ambitious objectives of the Vision into the 2005 NASA Authorization, directing NASA to return to the moon by 2020 to promote exploration, science, and commerce, and also to serve as a stepping-stone to Mars and other deep space destinations.[2] Each phase of exploration (Earth orbit, the Moon, and ultimately Mars) would build on the experience and lessons learned from earlier missions. Constellation hardware included Ares launch vehicles, an Earth Departure Stage secondary booster, an Orion spacecraft, and an Altair lunar lander.

In 2009, President Obama ordered a review of the Constellation program and acting Administrator Christopher Scolese established the "Review of U.S. Human Spaceflight Plans Committee", commonly referred to as the Augustine Commission. The Commission found that "since Constellation's inception the program has faced a mismatch between funding and program content" and that the funding strategy for Constellation relied on NASA retiring the Space Shuttle by 2010 and decommissioning the ISS by 2016.[3] The Commission proposed five alternative approaches for human space exploration, only two of which aligned with the Obama Administration's FY2010 budget profile for Constellation. Neither option would "permit human exploration to continue in any meaningful way", and ultimately the Obama Administration's FY2011 budget proposed cancellation of the Constellation program, shifting instead to an approach that would land humans on the surface of an asteroid.[4] While many elements of the Constellation program were abandoned, Congress directed NASA to develop a Space Launch System using "existing vehicle development and associated contracts" (i.e., efforts formerly dedicated to the development of the Ares launch vehicle) and the Orion spacecraft.[5]

---

[1] NASA, "The Vision for Space Exploration," Available at: https://www.nasa.gov/wp-content/uploads/2023/01/55583main_vision_space_exploration2.pdf.

[2] National Aeronautics and Space Administration Authorization Act of 2005, Pub. L. No. 109-115, 119 Stat. 2895 (2005).

[3] NASA, "Review of U.S. Human Spaceflight Plans Committee," Available at: https://www.nasa.gov/wp-content/uploads/2015/01/617036main_396093main_HSF_Cmte_FinalReport.pdf.

[4] John Logsdon, "A new US approach to human spaceflight?" Space Policy Vol. 1, issue 1, February 2011. Available at: https://www.sciencedirect.com/science/article/pii/S0265964610001189.

[5] National Aeronautics and Space Administration Authorization Act of 2010, Pub. L. No. 111-267, 124 Stat. 2805 (2010).

In 2017, Congress reiterated its continued support for the stepping-stone approach in the NASA Transition Authorization Act.[6] The Trump Administration was aligned on this position and issued Space Policy Directive-1 (SPD-1) directing NASA to "lead the return of humans to the Moon for long-term exploration and utilization."[7] This time, however, NASA would not go alone; per SPD- 1, the revived effort would involve a team of commercial and international partners.

In September of 2018, NASA issued its National Space Exploration Campaign Report describing NASA's efforts to plan a human lunar landing in the late 2020s.[8] This objective was confirmed six months later in NASA's FY2020 budget request, which announced NASA's intent to return humans to the Moon by 2028. Vice President Michael Pence further accelerated this deadline, directing NASA to land humans on the south pole of the Moon by 2024. NASA's FY2021 budget request reflected both the Artemis program and the 2024 landing date set by the Vice President. NASA also published its Lunar Exploration Program Overview in 2020, which provided an overview of the agency's planned lunar exploration activities.[9]

The Biden Administration has continued progress on the Artemis program's return to the Moon.[10] In 2022, Congress also required that NASA establish a new Moon to Mars Program Office within ESDMD, charged with ensuring that Artemis missions fit within the human exploration roadmap and facilitate a human mission to Mars.[11]

Artemis Elements

Artemis-related activities can be found in multiple NASA mission directorates. The primary branch responsible for Artemis elements is the

---

[6] National Aeronautics and Space Administration Transition Authorization Act of 2017, Pub. L. No. 115-10, 131 Stat. 18 (2017).

[7] Space Policy Directive-1, "Reinvigorating America's Human Space Exploration Program" (December 11, 2017). Available at: https://www.govinfo.gov/content/pkg/FR-2017-12-14/pdf/2017-27160.pdf.

[8] NASA, "National Space Exploration Campaign Report," Available at: https://www.nasa.gov/wp- content/uploads/2015/01/nationalspaceexplorationcampaign.pdf.

[9] NASA, "NASA's Lunar Exploration Program Overview," Available at: https://www.nasa.gov/wp- content/uploads/2020/12/artemis_plan-20200921.pdf.

[10] Elizabeth Howell, "US still committed to landing Artemis astronauts on the moon, White House says," Space News, February 5, 2021. Available at: https://www.space.com/biden-administration-commits-to-artemis-moon- landings.

[11] Chips and Science Act, Pub. L. No. 117-167 (2022).

Exploration System Development Mission Directorate (ESDMD), but the Space Technology Mission Directorate (STMD) and the Science Mission Directorate (SMD) also play key roles. The major elements of the Artemis program are set forth below.

Space Launch System (SLS): SLS is a two-stage, super heavy-lift launch vehicle operated at the Kennedy Space Center. Derived from the Constellation program's canceled Ares V launch vehicle, SLS uses RS-25 engines and solid rocket boosters adapted from the Shuttle program. NASA plans for three different SLS configurations:

- Block 1 (which includes a core stage, Interim Cryogenic Propulsion Stage (ICPS), and solid rocket boosters).
- Block 1B (which retains the core stage and solid rocket boosters, but replaces the ICPS with the Exploration Upper Stage (EUS)).
- Block 2 (retains the core stage and the EUS, but replaces the solid rocket boosters with an upgraded model).

Each configuration will result in greater SLS lift capacity, with the Block 2 capable of lifting 130 metric tons to Low Earth Orbit.

Orion Spacecraft (Orion): The Orion multipurpose spacecraft is a crew vehicle designed to carry astronauts between Earth and deep space. Orion can sustain a crew for up to 21 days of space exploration. For Artemis missions, Orion will carry crew from Earth to lunar orbit, and then from lunar orbit back to Earth (transport to the lunar surface and back will be provided by the Human Landing System discussed below). Orion consists of three main components: a crew module, a service module, and a launch abort system.

Exploration Ground Systems (EGS): EGS manages the development and operation of Kennedy Space Center systems and facilities that support modern and next generation launch vehicles and spacecraft. For Artemis, EGS is responsible for the capabilities used to assemble, launch, and recover SLS and Orion, which includes integration of the SLS and Orion systems in preparation for launch.

Gateway: Gateway is a small, multi-purpose space station that will be placed in lunar orbit to serve as both a staging point for lunar expeditions and deep space exploration, as well as a platform for scientific research and technology demonstrations. NASA intends for Gateway to be an international effort, and anticipates partners providing additional habitation modules, external robotics, refueling capabilities and other contributions. NASA already has established partnerships with international space agencies, including the European Space Agency, the Japan Aerospace Exploration Agency, and the Canadian Space Agency. NASA also recently announced a partnership with the UAE, under which the Mohammed bin Rashid Space Centre will "provide Gateway's Crew and Science Airlock module, as well as a UAE astronaut to fly to the lunar space station on a future Artemis mission."[12] The first four elements of Gateway are as follows:

- *Power and Propulsion Element (PPE):* PPE will provide power, thrust, and communications capabilities for Gateway.
- *Habitation and Logistics Outpost (HALO):* HALO provides basic habitation support infrastructure for Gateway, as well as additional docking ports for Orion and other spacecraft. HALO also can store cargo and other logistics deliveries that will support crewed missions.
- *International Habitat (I-Hab):* Like HALO, the I-Hab will provide additional spacecraft docking parts and living quarters for visiting astronauts. The I-Hab will be supplied by the European Space Agency (ESA).
- *ESPRIT Refueling Module (ERM):* ERM, also developed by ESA, will supply Gateway's propulsion system with fuel and also will provide additional storage space for cargo.
- *Human Landing System (HLS):* HLS will dock either with Gateway or Orion and will transport astronauts from lunar orbit to the surface of the Moon and back to lunar orbit. NASA awarded contracts to build landing systems to two United States commercial providers. SpaceX, selected in 2021, will develop an HLS based on its Starship spacecraft that will be used for Artemis III and IV.[13] Following direction from Congress, NASA opened another HLS solicitation[14] and picked Blue Origin as a secondary HLS provider in 2023.[15]

---

[12] NASA, "NASA, United Arab Emirates Announce Artemis Lunar Gateway Airlock," Available at: htps://www.nasa.gov/news-release/nasa-united-arab-emirates-announce-artemis-lunar-gateway-airlock/

[13] NASA, "NextSTEP H: Human Landing System," Available at: https://www.nasa.gov/humans-in-space/nextstep- h-human-landing-system/.

[14] NASA, "NASA Provides Update to Astronaut Moon Lander Plans Under Artemis," Available at: https://www.nasa.gov/news-release/nasa-provides-update-to-astronaut-moon-lander-plans-under-artemis/.

[15] NASA, "NASA Selects Blue Origin as Second Artemis Lunar Lander Provider," Available at: https://www.nasa.gov/news-release/nasa-selects-blue-origin-as-second-artemis-lunar-lander-provider/.

- *Space Suits:* NASA requires new spacesuits that are suitable for deep space environments, including the lunar surface. While NASA initially planned to produce the suits internally, the agency shifted its acquisition approach and instead opted for a commercial procurement.[16] In June of 2022, NASA awarded contracts to Axiom Space and Collins Aerospace to produce new suits via the Exploration Extravehicular Activity Services (xEVAS) program.[17]

Artemis Missions

The Artemis missions use the elements described above to access deep space destinations, including lunar orbit, Gateway, and/or the lunar surface. Each Artemis mission is distinguished by a different number.

Artemis I launched from the Kennedy Space Center on November 16, 2022. This mission originally was scheduled to launch in November of 2018, but experienced years of delays caused by SLS and Orion manufacturing complications, technical issues (including hydrogen leaks found during SLS wet dress rehearsals), and other programmatic challenges.[18]

The mission was an uncrewed demonstration mission and the first test of the fully integrated SLS, Orion, and EGS systems. During the 25-day mission, NASA tested the Orion spacecraft by performing two lunar flybys before returning to Earth on December 11, 2022. Upon return, NASA conducted post-flight analysis indicating that the mission was successful and many systems performed better than expected.[19]

Artemis II will be the first crewed demonstration mission of the integrated SLS, Orion, and EGS systems. Over the course of ten days, astronauts onboard Orion will confirm that all spacecraft systems operate as designed and test

---

[16] GAO, "NASA Lunar Programs: Moon Landing Plans Are Advancing but Challenges Remain," GAO-22-105533, March 1, 2022. Available at: https://www.gao.gov/products/gao-22-105533.

[17] Elizabeth Howell, "NASA just picked these 2 companies to build next-gen spacesuits for the moon, space station," Space News, June 1, 2022. Available at:https://www.space.com/nasa-selects-companies-build-spacesuits- moon-space-station.

[18] Carlyn Kranking, "Artemis 1 Launch Postponed Again and What Else You Need to Know About the Mission," Smithsonian Magazine, September 5, 2022. Available at: https://www.smithsonianmag.com/science-nature/what-you-need-to-know-about-nasas-artemis-i-launch-180980654/.

[19] NASA, "Analysis Confirms Successful Artemis I Moon Mission, Reviews Continue," Available at: https://www.nasa.gov/humans-in-space/analysis-confirms-successful-artemis-i-moon-mission-reviews-continue-2/

performance of the crewed spacecraft in deep space. The Artemis II crew includes three NASA astronauts (Reid Wiseman, Victor Glover, and Christina Koch) as well as an astronaut from the Canadian Space Agency (Jeremy Hansen). NASA's original baseline commitment was to launch Artemis II in April 2023. NASA now estimates Artemis II will launch in September of 2025.

Artemis III will be a crewed lunar landing demonstration mission. After launch, the crew's Orion spacecraft will travel to lunar orbit where it will rendezvous with SpaceX's Starship HLS. Once docked, two astronauts will board the Starship HLS, which will disconnect from Orion and descend to the lunar surface. Astronauts will spend approximately one week on the Moon, performing a range of tasks including scientific experiments and technology demonstrations. The Starship HLS will then transport the two astronauts back to lunar orbit to join their colleagues on Orion for return to Earth. NASA estimates that Artemis III will launch in September of 2026.

Artemis IV will be the first Artemis mission to utilize the SLS Block 1B configuration, which includes the EUS. Astronauts will travel onboard the Orion to lunar orbit, where they will deliver the I-Hab module to the Gateway. Then, two astronauts will board a Starship HLS and descend to the lunar service for a week of tasks, including collection of samples to bring back to Earth.

Artemis V, also using an SLS Block 1B, will deliver crew to lunar orbit and the ESPRIT module to Gateway. Two astronauts will again travel to the lunar surface to collect additional samples for return to Earth.

By the end of the 2020s, NASA intends to establish an SLS launch cadence of roughly one mission per year. NASA already is working to establish long-lead contracts to achieve this goal. For example, NASA has awarded a contract for the SLS solid rocket boosters that extends through Artemis XII.

Key Issues

The Artemis program has already seen both cost and schedule growth from its established baseline commitments. Despite forward progress on Artemis program initiatives, there are a number of risks that NASA must mitigate moving forward. Establishing an improved understanding of project cost and schedule, finalizing design and technical requirements, and resolving contract and personnel management concerns will all be important matters to consider moving forward. Artemis also faces difficulties stemming from the maturity of technologies critical to future missions. Below is a summary of key issues identified in recent reports, reviews, and audits of the Artemis program.

Government Accountability Office (GAO)

NASA Artemis Programs: Crewed Moon Landing Faces Multiple Challenges

GAO released a November 2023 report evaluating NASA's plan to complete a lunar landing on the Artemis III mission.[20] The report highlighted multiple challenges, including delays in the development of the lunar lander and spacesuits needed for the mission. The report concluded that a variety of factors, particularly the readiness of HLS, made a 2025 lunar landing unlikely. The challenges GAO identified include:

- An overly ambitious development schedule for the HLS program: GAO estimated that NASA's launch date was 13 months too short when compared to NASA's usual rate of production. If HLS development follows the average speed for major NASA projects, GAO estimated that HLS would not be ready for launch until early 2027.
- Delays in critical milestones: GAO found that 8 of 13 key events for the HLS program had been delayed by at least 6 months. SpaceX attempted a Starship Orbital Test Flight in April of 2023, but the flight was terminated early by the FTS system. Many subsequent tests are contingent on a successful Orbital Flight Test, causing strain on the already-compressed development timeline.
- Multiple novel and complex technical capabilities critical to the HLS design have yet to be matured: SpaceX must complete a large volume of complex technical work for HLS, especially in the areas of on-orbit propellant transfer and storage.

---

[20] GAO, "NASA Artemis Programs: Crewed Moon Landing Faces Multiple Challenges," GAO-24-106256, November 30, 2023. Available at: https://www.gao.gov/products/gao-24-106256.

NASA Lunar Programs: Improved Mission Guidance Needed as Artemis Complexity Grows

GAO released a September 2022 report assessing NASA's mission-level management for the Artemis program, including its development of mission schedules and mission-level reviews.[21] The report identified the following concerns:

- NASA lacks "agency-wide, mission-level schedule management guidance to inform realistic integration schedules and launch dates for Artemis missions." NASA instead adapts guidance that was developed for program-level schedule management rather than mission-level.
- NASA has yet to conduct a Schedule Risk Analysis (SRA) for Artemis II.
- NASA has not developed a mission-level schedule for Artemis III.
- While NASA conducts workforce planning, it does not perform any advance workforce planning beyond five budget years. NASA already has committed billions of dollars for Artemis contracts that extend well beyond this five-year window. NASA risks facing a shortage of skilled laborers needed for future Artemis activities.

GAO recommended that NASA:

- Direct the NASA Chief Financial Officer to coordinate with mission directorates for development of mission-level schedule management guidance for Artemis.
- Conduct a schedule risk analysis for the Artemis II mission and update it as needed to incorporate schedule updates and new risks.
- Develop guidance for division-level schedule collaboration on Artemis III and subsequent missions.
- Ensure that the NASA Office of the Chief Human Capital Officer develops guidance identifying a regular and recurring process for long-term Artemis workforce scenario planning at least 5 years beyond the existing 5-year workforce plans.

---

[21] GAO, "NASA Lunar Programs: Improved Mission Guidance Needed as Artemis Complexity Grows," GAO-22-105323, September 8, 2022. Available at: https://www.gao.gov/products/gao-22-105323.

Other Reports

GAO has released several reports regarding the Artemis program. In an analysis of SLS cost transparency, GAO stated:

> "NASA does not plan to measure production costs to monitor the affordability of the SLS program. After SLS's first launch, Artemis I in November 2022, NASA plans to spend billions of dollars to continue producing multiple SLS components, such as core stages and rocket engines, needed for future Artemis missions. These ongoing production costs to support the SLS program for Artemis missions are not captured in a cost baseline, which limits transparency and efforts to monitor the program's long-term affordability."[22]

When reviewing programmatic challenges of Artemis I through III, GAO noted that, due to the sequential links between each of the first three missions, delays to one mission will have cascading cost and schedule impacts for the other missions.[23] Further, the minimum time required between Artimis I and II, and Artemis II and III limits NASA's ability to mitigate the effects of these delays. GAO also highlighted a noticeable lack of cost and schedule baselines for many Artemis projects, which creates challenges in assessing the progress and affordability of the program. For example, Orion does not have a cost and schedule baseline past Artemis II.

NASA Inspector General

NASA's Management of the Artemis Supply Chain

NASA OIG issued a report in October of reviewing NASA's management of the Artemis supply chain and analyzing problems.[24] NASA IG noted that, while many of the challenges it identified were outside of NASA's control, "the Agency lacks visibility into its critical suppliers with many Artemis programs and projects not tracking their prime contractors' supply chain

---

[22] GAO, "Space Launch System: Cost Transparency Needed to Monitor Program Affordability," GAO-23-105609, September 7, 2023. Available at: https://www.gao.gov/products/gao-23-105609.

[23] *Id.* at 16.

[24] NASA OIG, "NASA's Management of the Artemis Supply Chain," IG-24-003, October 19, 2023. Available at: https://oig.nasa.gov/docs/IG-24-003.pdf.

impacts." Additionally, the IG found that Artemis programs and projects were not taking advantage of NASA's Logistics Management Division (LMD) when addressing supply chain issues. More generally, the report noted that NASA's project management practices fell short of other government agencies conducting major projects. It also concluded that NASA's efforts to improve supply chain visibility thus far have been ineffective.

The NASA IG provided many recommendations for NASA, including suggestions that NASA:

- Provide training and resources to ensure that contracting officers utilize available supplier data.
- Centralize supply chain management for the Artemis campaign within the Moon to Mars Program Office.
- Incorporate a representative from LMD into each Artemis-related program.
- Ensure an Artemis-specific industrial base and supply chain study is completed on a recurring basis.

NASA's Partnerships with International Space Agencies for the Artemis Campaign

In January of 2023, NASA OIG issued a report assessing NASA's plans for international cooperation and identifying impediments to execution of international partnerships.[25] The report found:

- NASA lacks a comprehensive, overarching strategy to coordinate international contributions for the Artemis program.
- NASA lacks comprehensive forums (e.g., boards, panels, and working groups) to facilitate Artemis-related discussions with international partners.
- U.S. export control regulations present an obstacle, as such regulations "can be overly complex and restrictive, and their implementation in international agreements, policies, and how space flight systems are classified routinely limit NASA's international

---

[25] NASA OIG, "NASA's Partnerships with International Space Agencies for the Artemis Campaign," IG-23-004, January 17, 2023. Available at: https://oig.nasa.gov/docs/IG-23-004.pdf.

collaborations on Artemis." Further, "the Artemis campaign lacks a unique EAR classification of specific space flight items or consistent jurisdiction and classification of Artemis elements, such as the Orion spacecraft, that would simplify the timely exchange of space flight items and technical information with international partners."

Select recommendations from the report suggest that NASA leadership:

- Establish NASA-led Artemis campaign boards and working groups for partners with agreed-upon commitments and provide opportunities for liaison representation from international partner agencies.
- Perform a detailed gap analysis and cost estimate for Artemis missions beyond Artemis IV that will help inform a cost-sharing strategy with international partners.
- Review export control requirements and consider additional roles for partner astronauts to increase their utilization in NASA space flight operations.
- Execute Artemis agreements with key international space agency partners to ensure partner roles and responsibilities are clearly understood and allow for efficient and timely partnerships in support of Artemis.

NASA's Management of the Space Launch System Booster and Engine Contracts

Issued in May of 2023, this NASA OIG report explored performance of the Boosters and Adaptation contracts and reviewed the impact of Booster Production and Operations Contract (BPOC) and R-25 Restart and Production efforts to improve Artemis program cost management.[26] The IG found the following:

- NASA continues to face substantial cost growth, and schedule delays in the Artemis program that could impact technology design. Despite

[26] NASA OIG, "NASA's Management of the Space Launch System Booster and Engine Contracts," IG-23-015, May 25, 2023. Available at: https://oig.nasa.gov/docs/IG-23-015.pdf.

this, NASA concurrently is developing and producing engines and boosters. This conflicts with the established best practice of completing development before moving to production.

- Marshall Space Flight Center (MSFC) procurement officials charged with overseeing all four Artemis program contracts "are challenged by inadequate staff, their lack of experience, and limited opportunities to review contract documentation."
- NASA opted to use cost-plus contracts for projects where fixed-price contracts could potentially have reduced costs, including for added production engines under the RS-25 Restart and Production contract and acquisition of long-lead materials under the BPOC letter contract.

Select recommendations from the report suggest that NASA leadership:

- Assess whether the 18 new production engines under the RS-25 Restart and Production contract can be acquired through a fixed-price contract.
- Identify procurement needs and resources available to address MSFC staff shortages, and ensure that MSFC officials comply with best practices for establishing and maintaining internal controls related to requests for equitable adjustment of award fee payments, fiscal law, and appropriate internal and external engagement.
- Update the cost-per-engine estimate for RS-25 engines to include investments made in production restart.
- Develop a separate non-fee bearing contract line item for completion of the unfinished adaptation of heritage RS-25 engines.

Figures

The following figures provide additional information on the Artemis program, including budget estimates, program elements, and mission profiles.

Budget Charts

Artemis Campaign Development

Budget Authority (in $ millions)	Op Plan FY 2022	Enacted FY 2023	Request FY 2024	FY 2025	FY 2026	FY 2027	FY 2028
Gateway	742.5	--	**914.2**	853.0	744.2	768.8	777.3
Adv Cislunar and Surface Capabilities	70.1	--	**60.3**	102.0	433.0	563.8	969.9
Human Landing System	1,195.0	1,485.6	**1,880.5**	2,224.7	2,286.7	2,748.3	2,526.6
xEVA and Human Surface Mobility Program	0.0	275.9	**379.9**	494.8	605.0	605.3	605.7
Total Budget	**2,007.6**	**2,600.3**	**3,234.8**	**3,674.4**	**4,068.9**	**4,686.2**	**4,879.6**

Source: NASA FY2024 budget request.

Figure 1. NASA budget request for Artemis Campaign Development for FY2024 to FY2028

Artemis Program Components

Budget Authority (in $ millions)	Op Plan FY 2022	Enacted FY 2023	Request FY 2024	FY 2025	FY 2026	FY 2027	FY 2028
Orion Program	1,401.7	1,338.7	**1,225.0**	1,093.7	1,093.7	1,094.2	1,115.1
Crew Vehicle Development	1,388.8	1,320.3	**1,212.6**	1,058.7	1,058.7	1,058.5	1,062.5
Orion Program Integration and Support	12.9	--	**12.5**	34.9	35.0	35.7	52.7
Space Launch System	2,600.0	2,600.0	**2,506.1**	2,483.3	2,322.4	1,917.1	1,969.1
Launch Vehicle Development	2,526.9	2,361.4	**2,427.2**	2,365.8	2,206.7	1,804.6	1,798.8
SLS Program Integration and Support	73.1	--	**78.9**	117.5	115.7	112.5	170.3
Exploration Ground Systems	589.0	799.2	**794.2**	664.7	593.2	546.0	445.5
Exploration Ground Systems Development	398.1	330.6	**273.2**	143.5	81.8	15.6	0.0
EGS Program Integration and Support	190.9	--	**521.0**	521.2	511.4	530.4	445.5
Construction & Envrmtl Compl Restoration	90.3	--	**10.5**	0.0	0.0	0.0	0.0
Exploration CoF	90.3	--	**10.5**	0.0	0.0	0.0	0.0
Total Budget	**4,681.0**	**4,824.1**	**4,535.9**	**4,241.7**	**4,009.3**	**3,557.3**	**3,529.7**

source: NASA FY2024 budget request.

Figure 2. NASA budget request for Orion, SLS, and EGS for FY2024 to FY2028.

Budget Authority (in $ millions)	Op Plan FY 2022	Enacted FY 2023	Request FY 2024	FY 2025	FY 2026	FY 2027	FY 2028
Total Budget	**0.0**	**--**	**49.1**	**50.0**	**50.5**	**51.0**	**51.1**

source: NASA FY2024 budget request.

Figure 3. NASA budget request for Moon to Mars Architecture for FY2024 to FY2028.

Budget Authority (in $ millions)	Op Plan FY 2022	Enacted FY 2023	Request FY 2024	FY 2025	FY 2026	FY 2027	FY 2028
Total Budget	187.4	--	161.8	164.4	164.4	164.5	167.8

source: NASA FY2024 budget request.

Figure 4. NASA budget request for Mars Campaign Development for FY2024 to FY2028.

Artemis Operational Costs

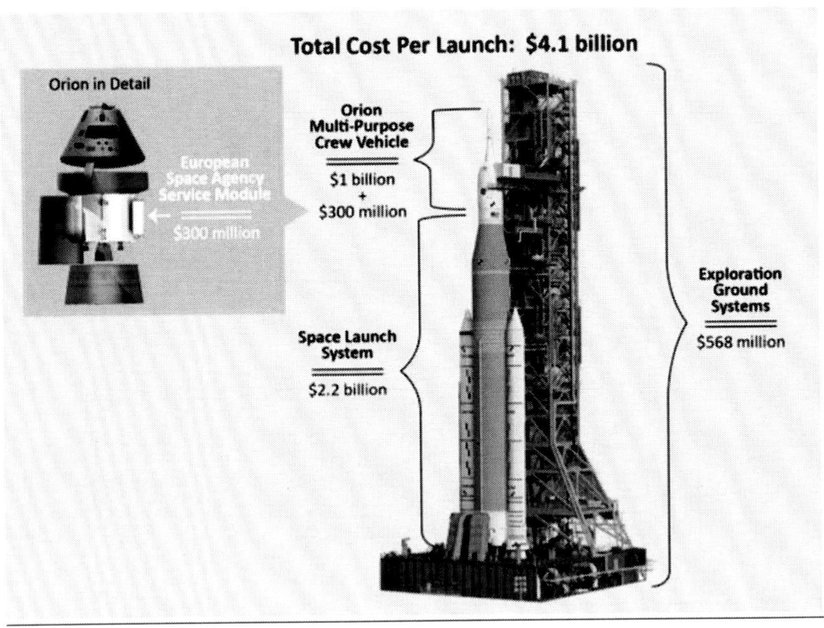

Source: NASA OIG.

Figure 5. NASA estimate of SLS and Orion Operating Costs Per Launch.

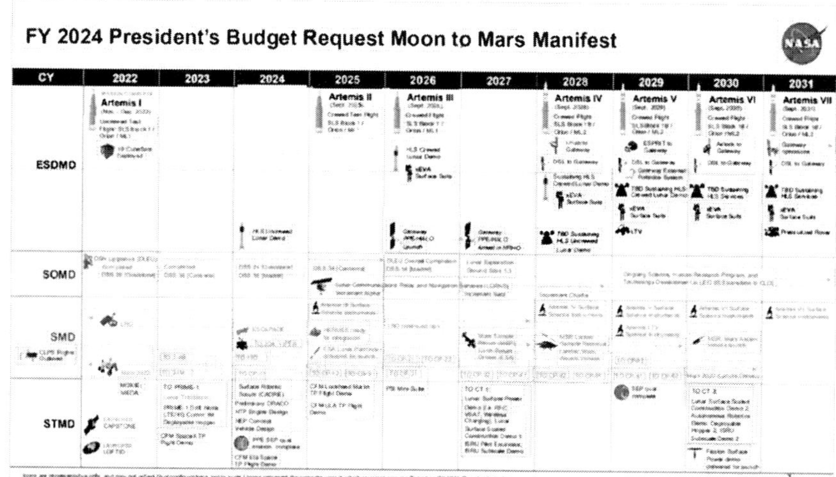

Source: NASA.

Figure 6. NASA schedule for Artemis and associated missions.

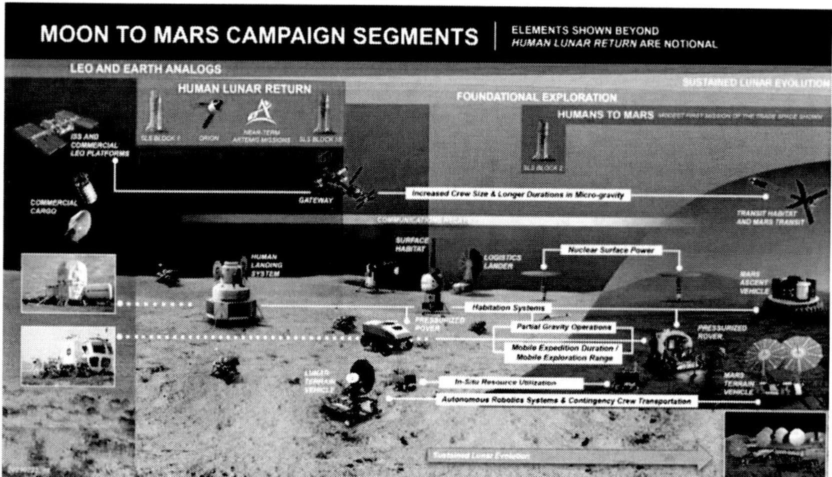

Source: NASA.

Figure 7. NASA Architecture for human deep space exploration.

Artemis Components

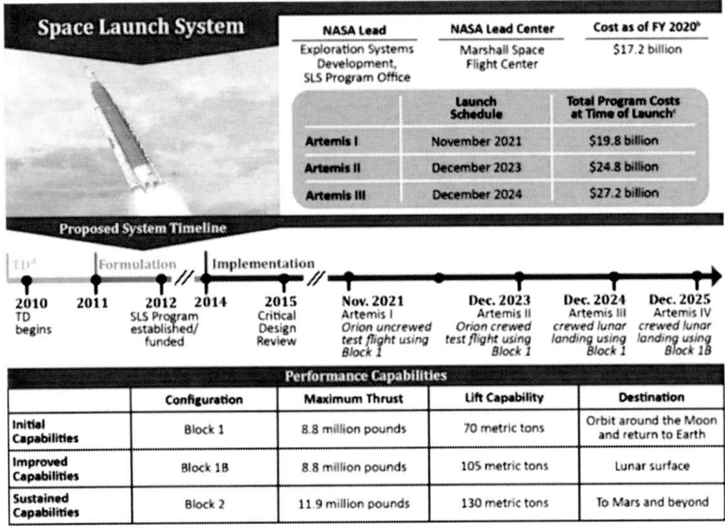

	Launch Schedule	Total Program Costs at Time of Launch[c]
Artemis I	November 2021	$19.8 billion
Artemis II	December 2023	$24.8 billion
Artemis III	December 2024	$27.2 billion

	Configuration	Maximum Thrust	Lift Capability	Destination
Initial Capabilities	Block 1	8.8 million pounds	70 metric tons	Orbit around the Moon and return to Earth
Improved Capabilities	Block 1B	8.8 million pounds	105 metric tons	Lunar surface
Sustained Capabilities	Block 2	11.9 million pounds	130 metric tons	To Mars and beyond

Source: NASA OIG.

Figure 8. SLS overview and development plan.

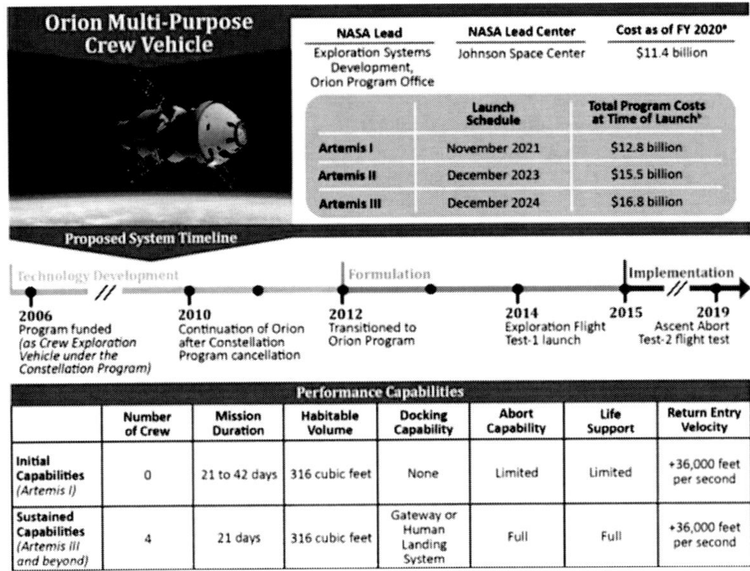

	Launch Schedule	Total Program Costs at Time of Launch[b]
Artemis I	November 2021	$12.8 billion
Artemis II	December 2023	$15.5 billion
Artemis III	December 2024	$16.8 billion

	Number of Crew	Mission Duration	Habitable Volume	Docking Capability	Abort Capability	Life Support	Return Entry Velocity
Initial Capabilities *(Artemis I)*	0	21 to 42 days	316 cubic feet	None	Limited	Limited	+36,000 feet per second
Sustained Capabilities *(Artemis III and beyond)*	4	21 days	316 cubic feet	Gateway or Human Landing System	Full	Full	+36,000 feet per second

Source: NASA OIG.

Figure 9. Orion overview and development plan. .

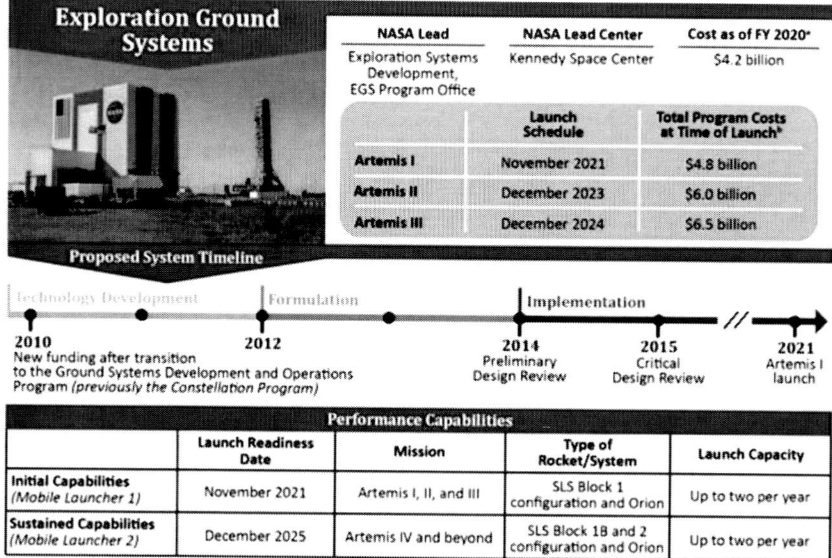

Source: NASA OIG.

Figure 10. EGS overview and development plan.

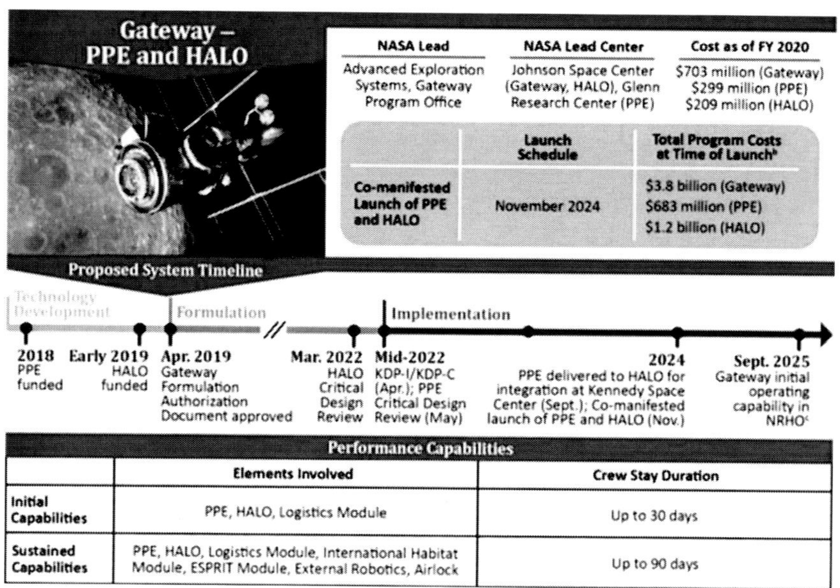

Source: NASA OIG.

Figure 11. Gateway overview and development plan.

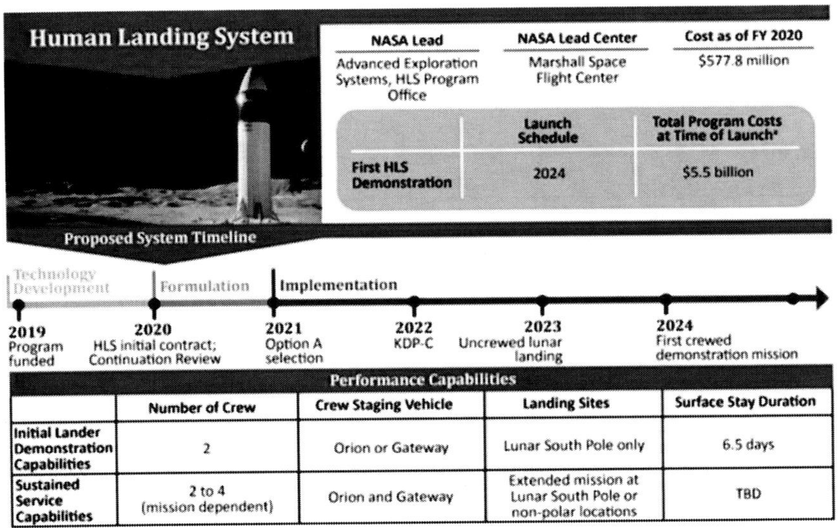

Source: NASA OIG.

Figure 12. HLS overview and development plan.

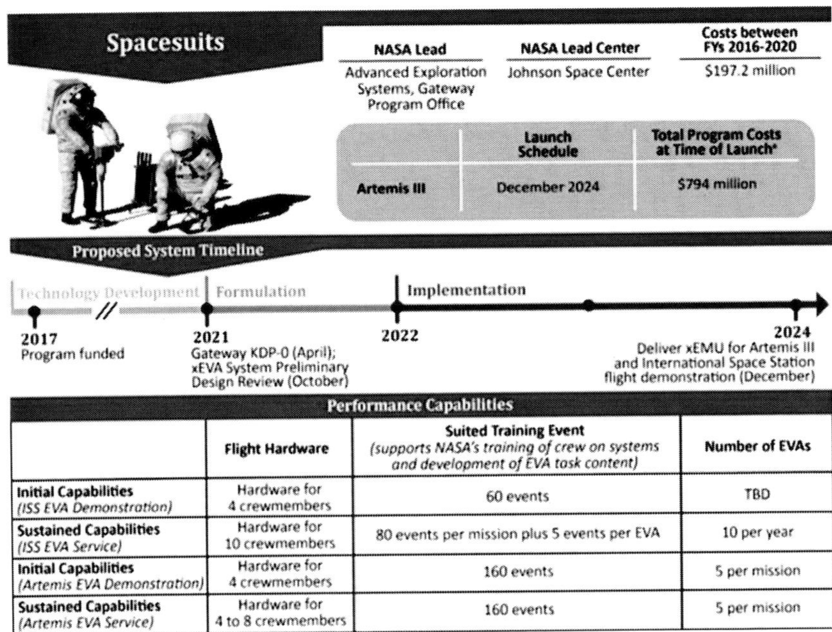

Source: NASA OIG.

Figure 13. Spacesuits overview and development plan.

Opening Statement of Chairman Frank Lucas Space and Aeronautics Subcommittee Hearing Returning to the Moon: Keeping Artemis on Track, January 17, 2024

Good morning. I want to welcome everyone to the Science Committee's first hearing of 2024. It's fitting that we're kicking off the year with a hearing on Artemis, given its importance to our space program and to U.S. competitiveness.

My top priority since becoming chairman of the Science Committee has been to ensure American competitiveness and leadership in the fields of research and technology development. This includes U.S. activities in space, especially human exploration.

The importance of U.S. leadership is space is why some of our top legislative priorities this Congress include a NASA authorization bill, which we will consider this spring, and the Commercial Space Act.

It has been almost seven years since a comprehensive NASA authorization bill was signed into law, and that is simply too long for an agency of NASA's importance. Much has happened during that period, and this Committee should provide direction to NASA's activities for the coming years, especially in the area of human exploration.

How we address future human exploration beyond Low Earth orbit is undoubtedly a topic we will address in the NASA authorization bill. Artemis is a cornerstone of that effort.

I am confident that I speak for everyone on this committee when I say we all support Artemis. This committee has long directed NASA to return humans to the Moon and eventually Mars.

But this Committee's support of Artemis means asking detailed questions of NASA and providing oversight of the agency's proposals. Congress must have proper insight into the agency's planning and execution of this mission to ensure its success.

This also means listening to inputs from external stakeholders and hearing differing viewpoints, which is why we have assembled a panel of witnesses with a variety of perspectives today.

Last week, NASA announced the delay of Artemis 2 to September 2025 and Artemis 3 to September 2026. I look forward to hearing from NASA about the cause of these delays and potential impacts to future missions, and about the steps it is taking to mitigate future risks.

We have a responsibility to not only our constituents, but the international community to see that Artemis is executed in a timely and fiscally responsible manner without sacrificing safety.

I remind my colleagues that we are not the only country interested in sending humans to the Moon. The Chinese Communist Party is actively soliciting international partners for a lunar research station and has stated its ambition to have astronauts on the human surface by 2030.

The country that lands first will have the ability to set a precedent for whether future lunar activities are conducted with openness and transparency or in a more restricted manner.

I am grateful to our panel for appearing before us today to share their experience and expertise. I look forward to a productive discussion on how we can ensure the success of Artemis and the best way for the U.S. to be the world leader in human space exploration.

I now recognize Ranking Member Sorensen for his opening statement.

Opening Stgatement of Ranking Member Eric Sorensen (D-IL) of the Subcomittee on Space and Aeronautics Space & Aeronautics Subcommittee Hearing: Returning to the Moon: Keeping Artemis on Track, January 17, 2024

Good morning and thank you, Chairman Lucas, for holding today's hearing Returning to the Moon: Keeping Artemis on Track.

I want to welcome our distinguished witnesses. Thank you for being here.

I was not alive during the Apollo 11 landing, but as the son of an aerospace engineer, and a meteorologist with a deep love of science, I know the profound impact it has had on our country and on the world. When I look up at the night sky, I wonder what is up there? I want us to go so I can know.

Today, we are examining NASA's Artemis program. This program, separated into several stages, is designed to bring humans, step by step, to the moon and beyond.

Artemis will inspire the next generation, strengthen our aerospace industry and international partnerships, and demonstrate capabilities needed to eventually send humans to Mars.

Last year, I was proud to host NASA astronaut, Dr. Kate Rubins, in my district. Dr. Rubins spoke about her excitement for the upcoming generation.

She believes that first graders are the perfect age to one day go to Mars. What an exciting possibility for our nation's children!

The Artemis I mission was an important first uncrewed test and sent the Orion vehicle thousands of miles beyond the Moon before its return to Earth. Artemis II will test additional systems as it brings humans around the moon. And Artemis III will land humans back on the moon. The difficulty of these missions cannot be underestimated.

Last week, we learned that NASA is delaying the Artemis II and Artemis III missions by about a year. I stand behind NASA in prioritizing safety for Artemis, and I look forward to gaining further insight into the delays and any related costs.

Artemis requires a sustained national investment. In a 2021 report, the NASA Office of Inspector General said, "NASA is projected to spend $93 billion on the Artemis effort from FY 2012 through FY 2025." And that's even before we land our astronauts on the Moon.

As authorizers with oversight responsibility, this committee needs to ensure those investments are made wisely.

This hearing provides a timely opportunity to get both an update on the progress and an understanding of the pressing issues for the Artemis program, including,

- Do NASA and Congress have an appropriate level of understanding of the cost of key Artemis systems, individual Artemis missions, and a sustained lunar exploration effort?
- What is on the critical path for returning humans to the Moon and what is the plan for addressing those challenges?
- How would an FY2024 budget at enacted FY2023 levels, or even a cut below the FY 2023 appropriated levels, affect the Artemis program?
- How are NASA and its partners addressing risks and how will risk be communicated to the public?

In closing, Mr. Chairman, I want Artemis to be safe and successful. Artemis and Moon to Mars are of tremendous importance to the United States and the world.

America's international leadership and engagement in the Artemis program and the Artemis accords will promote peaceful, safe, and sustainable exploration of the Moon and other celestial bodies.

Thank you, and I yield back the balance of my time.

Opening Statement of Ranking Member Zoe Lofgren (D-CA) Space & Aeronautics Subcommittee Hearing: Returning to the Moon: Keeping Artemis on Track, January 17, 2024

Good morning, and thank you, Chairman Babin, for holding today's hearing. I also want to welcome our witnesses. Thank you for being here to discuss the topic of "Returning to the Moon: Keeping Artemis on Track."

The Committee has long maintained its bipartisan support for Artemis and NASA's Moon to Mars efforts, and I don't see that changing in any way. I was thrilled with the success of the Artemis I test flight. In my own state of California, NASA's Moon to Mars campaign supports 11,600 jobs and created an economic impact of 2.8 billion dollars, according to NASA's FY2021 Economic Impact Report.

So let me be clear upfront. I support Artemis. I want it to be successful, especially with China at our heels. We want to be helpful in ensuring Artemis is strong and staying on track as we look to lead the world, hand in hand with our partners, in the human exploration of the Moon and beyond.

Sending people into space, let alone to the Moon, will never be easy. NASA recently announced delays to the Artemis II and III missions. I have full confidence in NASA's workforce and the decision to keep safety as the top priority. To that end, I look forward to understanding the details behind the recent delays and what is involved in addressing the issues.

As Artemis efforts continue, it's incumbent upon us, as the authorizing committee, to have our eyes wide open. Moon to Mars is a multi-decadal effort that will span several Congresses and Administrations.

Full situational awareness requires that:

1. We know how much the key Artemis systems cost, as well as the missions themselves;
2. Have a realistic understanding of how NASA is assessing schedule;
3. Have clarity on the topmost technical challenges and risks and how they are being addressed across NASA and among its diverse set of partners and acquisition mechanisms.

Moreover, we can't ignore that NASA has a lot on its plate. The future of low Earth orbit and the planned end of International Space Station operations in 2030, the need for a critical yet costly deorbit vehicle, the transition to the use of future commercial space stations and their readiness to come online must be kept in mind. In addition, key considerations on Mars Sample Return are on the horizon. And, as we learned last week from NASA and NOAA's annual assessment of global temperature, we must continue to obtain the measurements and observations needed to understand and mitigate the horrific impacts of the climate crisis. In short, NASA is a multi- mission agency, and we can't lose sight of the benefits and challenges of a balanced portfolio.

Supporting balance won't be made any easier by the dysfunctional appropriations process that threatens to undermine what we know is best for leading the world and growing our economy in a sustainable way—investments in R&D and innovation such as those at NASA.

I'm excited about Artemis and Moon to Mars. I look forward to working with the Chairman, the Administration, and stakeholders on building a smart, strong, and sustainable path forward.

Thank you, and I yield back.

House Science, Space, and Technology Committee Space and Aeronautics Subcommittee Hearing titled "Returning to the Moon: Keeping Artemis on Track" Statement by Subcommittee Chairman Brian Babin Wednesday, January 17, 2024

Good morning. I want to welcome everyone to the Science Committee's first hearing of 2024. It's fitting that we're kicking off the year with a hearing on Artemis, given its importance to our space program and to U.S. competitiveness.

The nation that leads in space earns tremendous scientific knowledge, reaps the rewards of technological advancements, and sets the rules of the road for future exploration. It's critical that we continue to lead so that our values of transparency, openness, and freedom guide exploration rather than communist principles and dictatorial regimes. That's why it's so important for Artemis to succeed.

The origins of the Artemis program stem from President Bush's Vision for Space Exploration, announced in January of 2004. In 2005, this committee directed NASA to plan to return American astronauts to the Moon as a stepping-stone to Mars and beyond. This committee, and Congress as a whole, has not wavered in its commitment to that goal. All too often NASA programs have suffered from cost over-runs, under-performance, schedule delays, or changing political directions that have led to cancellations. Recognizing this history, Congress has provided "continuity of purpose" for Artemis through multiple NASA Authorization Acts, robust appropriations, and consistent oversight to ensure the program remained focused across several Administrations.

This was no small task, and we still have our work cut out for us to maintain the program and ensure success. I was incredibly pleased to see the success of Artemis' first mission in November of 2022, which sent an uncrewed Orion capsule around the Moon and back to Earth, where it was successfully recovered in the Pacific Ocean. But last week, NASA announced delays to the Artemis 2 mission, which would send astronauts around the Moon, and the Artemis 3 mission, which would return humans to the lunar surface for the first time in more than 50 years. Artemis 2 has been delayed until September 2025 and Artemis 3 has been pushed back to September 2026. This is in addition to proposed delays to Artemis 4 that were included in the President's Fiscal Year 2024 budget request last year. While an argument could be made that those schedules were aggressive, it is important for Congress to monitor contract performance and NASA program management to gain insight into trends and indicators that could portend future issues.

Every delay costs the United States time and taxpayer dollars and risks our preeminent role in space exploration. As I said at the beginning of my remarks, we cannot afford to cede U.S. leadership in space, so it's critical that we keep Artemis on track and on time.

That is the focus of the hearing today. My goal is for this Committee to come away with a better understanding of the current challenges facing Artemis and our efforts to return to the Moon.

There are plenty of topics for us to explore today ranging from acquisition strategies, architecture decisions, concept of operation choices, contractor performance, and NASA oversight. While we will only touch the surface of these complicated issues today, we will surely continue our oversight through additional hearings, information requests, budget reviews, and stakeholder engagement.

Today, however, we have witnesses from NASA, the Government Accountability Office (GAO), the NASA Inspector General, and the private sector, all of whom can give us more insight into the program and what's needed to keep it moving forward on time and on budget.

I look forward to their testimony, and discussing how we can ensure future success. Thank you.

Committee on Science, Space and Technology Subcommittee on Space and Aeronautics, U.S. House of Representatives Statement by Catherine Koerner, Associate Administrator, Exploration Systems Development Mission Directorate, NASA before the Subcommittee on Space & Aeronautics, House Science, Space & Technology

Chairman Babin, Ranking Member Sorensen and distinguished members of the Subcommittee, thank you for the opportunity to testify on NASA's Artemis program. Under the Artemis program, the United States, along with our international and commercial partners, will return humans to the Moon to explore, conduct scientific research, and establish the capability for a long-term human presence on and around the Moon. Then, using what we learn at the Moon, we will take the next giant leap: sending the first humans to Mars.

In November 2022, NASA took the first major step in America's return to the Moon with the Artemis I mission. That historic launch and 25 and a half-day mission tested the Space Launch System (SLS) rocket, the Orion spacecraft, and the Exploration Ground Systems in preparation for Artemis II. On Artemis II, NASA Astronauts Reid Wiseman, Victor Glover, Christina Koch, and Canadian Astronaut Jeremy Hanson will journey beyond low-Earth orbit and around the Moon – the farthest humans have journeyed into space in more than 50 years. Approximately one year after Artemis II, the Artemis III crew will land at the lunar South Pole and begin building out a robust, long-term lunar exploration program. With Artemis IV, astronauts will again visit the lunar surface and start assembly of a space station in lunar orbit called the Gateway.

NASA's plan for a successful and sustainable return to the Moon requires the development of several new space systems, including the SLS rocket, the Orion spacecraft, the Exploration Ground Systems, lunar landers, the Gateway space station, and new lunar spacesuits and lunar rovers. Last year, pursuant to the NASA Authorization Act of 2022, NASA established the Moon to Mars program office, which focuses on development of these new systems, mission integration, and risk management across the portfolio. This new office also leads planning and analysis for long-lead technology developments to support human Mars missions.

In the year since NASA's successful Artemis I flight test around the Moon, NASA has continued to refine the schedule of the follow-on Artemis missions, based on data from the Artemis I mission and the readiness of the space systems needed to safely transport our crews from Earth to the lunar surface and back. Artemis II adds several new systems to support astronauts inside Orion. In addition, we are continuing to study the Orion heat shield from Artemis I, to ensure the safety of our crew on future missions. Based on these factors, we are planning for Artemis II to launch in September 2025. Artemis III will build on the progress of Artemis I and II and adds a commercial lunar lander and advanced spacesuits for walking on the lunar surface. In 2026, Artemis III will send humans back to the surface of the Moon.

While sending humans back to the Moon will be a significant accomplishment, we do not intend to stop there. NASA's long-term goal is to send humans to Mars – and the Moon will help us get there. Mars is a rich destination for scientific discovery and a driver of technologies that will enable humans to travel and explore far from Earth. By using what we learn on and around the Moon under Artemis, NASA is working to understand and overcome the future challenges associated with landing and living on Mars. As NASA builds a blueprint for human exploration throughout the solar system for the benefit of humanity, we conducted our first two Architecture Concept Reviews, the culmination of a robust analysis process designed to align NASA's Moon to Mars exploration strategy and codify the supporting architecture. This annual review is a milestone that enables our Moon to Mars strategy to evolve over time as we consider lessons learned from previous missions and provide opportunities to on-ramp new technologies as well as new industry and international partners.

Through the Artemis campaign, NASA is partnering with the most diverse and broadest exploration coalition in history, including multiple international and commercial partners. For example, NASA's Gateway Program is an international collaboration with the Canadian Space Agency, European Space Agency, Japan Aerospace Exploration Agency, and now the United Arab Emirates' Mohammed bin Rashid Space Centre, to establish humanity's first space station around the Moon. Similarly, NASA is exploring additional international partnerships for lunar surface habitats, logistics, and mobility capabilities that will enable long-term human presence and enhanced science returns.

Together, we will continue to develop the technology and systems needed to live and work on and around the Moon in preparation for human missions to Mars. Because of our diverse astronaut corps, we will enable the first woman, the first person of color, and the first international astronaut to walk on the Moon. We will align with partners toward a future of expanded economic opportunity and scientific discovery, while investing in the next generation of STEM leaders as we support the limitless possibilities of space exploration.

NASA is grateful for this committee's continued support of the Artemis program, and I appreciate this opportunity to update you on behalf of NASA and our Artemis partners and would be pleased to answer your questions.

Catherine Koerner
Associate Administrator for Exploration Systems Development

Catherine Koerner is the associate administrator for the Exploration Systems Development Mission Directorate at NASA Headquarters in Washington. She is responsible for the development of NASA's Moon to Mars architecture, defining and managing the systems development for Artemis missions, and planning for integrated deep space exploration approach.

Koerner was formerly the deputy associate administrator for the mission directorate, providing leadership and management of human spaceflight development and operations related to NASA's Moon and Mars exploration goals. She was responsible for establishing and defining future space exploration architectures while overseeing development of new space transportation systems and supporting capabilities that are critical for human-led deep space exploration and scientific research.

Prior to her position at NASA Headquarters, Koerner was NASA's Orion Program manager at NASA's Johnson Space Center in Houston, where she was responsible for oversight of design, development, and testing of the Orion spacecraft. Before leading the Orion Program, Koerner served as the director of Human Health and Performance Directorate, focusing on enhancing crew health and performance and mitigating risks associated with human spaceflight. The core capabilities in the directorate include space and clinical operations; biomedical research and environmental sciences; human systems engineering and development; strategic planning, benchmarking, collaboration, and open innovation.

As a former NASA flight director at NASA Johnson, Koerner led teams in mission control during space shuttle and International Space Station missions. She also previously held several leadership positions within the space station program during its assembly phase and managed NASA's cargo resupply services contracts for it, helping foster a commercial space industry in low Earth orbit. Before joining Johnson in 1991, she worked at NASA's Jet Propulsion Laboratory in Southern California, where her career with the agency began as a mission design engineer.

Koerner earned her Bachelor of Science and Master of Science degrees in aeronautical and astronautical engineering from the University of Illinois at Urbana-Champaign. She has received numerous awards including a Presidential Rank Award in 2019, two Outstanding Leadership Medals (2006, 2013), NASA's Exceptional Service Medal (2007), Johnson's Center Director Commendation (2017) and numerous Group Achievement Awards.

United States Government Accountability Office, Testimony Before the Subcommittee on Space and Aeronautics, Committee on Science, Space, and Technology, House of Representatives

NASA Artemis Programs: Lunar Landing Plans Are Progressing, but Challenges Remain

Statement of William Russell, Director, Contracting and National Security Acquisitions

Why GAO Did This Study

The National Aeronautics and Space Administration (NASA) is committing billions of dollars to return humans to the lunar surface and initiate human exploration of Mars. The missions, known collectively as Artemis, involve the development and integration of multiple systems and programs.

This testimony focuses on NASA's progress toward achieving the Artemis missions, as well as the challenges the agency faces in conducting them. This statement is based on past GAO reports on the Artemis enterprise and our annual assessment of NASA's major projects.

What GAO Recommends

GAO has made numerous recommendations to reduce NASA's acquisition risk and improve NASA's management of its portfolio of major projects, which includes projects supporting the Artemis missions.

NASA has generally agreed with the recommendations and implemented changes in response to many of them. However, it needs to take additional actions to fully address all of them.

As of May 2023, GAO identified six open recommendations related to Artemis as being a priority for implementation. For example, NASA concurred with our December 2019 recommendation to create a life-cycle cost estimate for the Artemis III mission but has not yet implemented it.

What GAO Found

NASA has made progress demonstrating key capabilities needed to support its Artemis missions:

- Artemis I, an uncrewed test flight, successfully launched in November 2022, which demonstrated the initial capability of the Space Launch System and Exploration Ground Systems.
- For Artemis II, the first flight with crew, NASA is currently conducting integration and testing of the crew capsule and the launch pad.
- NASA and its contractors continue to make progress on technologies supporting Artemis III, the first crewed lunar landing mission. For example, the human landing system contractor has conducted two test flights of its human landing system.

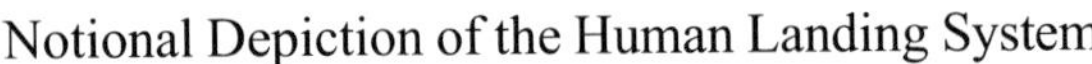
Notional Depiction of the Human Landing System

Source: SpaceX. | GAO-24-107249.

Despite this progress, NASA still faces several challenges:

- *Ambitious schedules.* In November 2023 (GAO-24-106256), GAO found that the Artemis III lunar landing was unlikely to occur in December 2025, as planned, given delays and remaining technical work. In January 2024, NASA adjusted the launch date to September 2026 to allow contractors time to complete a significant amount of remaining complex work.

- *Artemis III mission cost.* In December 2019 (GAO-20-68), GAO found that NASA did not plan to establish an official cost estimate for this mission. NASA concurred with a GAO recommendation to establish one but has not yet done so. While NASA requested $6.8 billion to support Artemis III programs in its fiscal year 2024 budget request, decision-makers have limited knowledge into the full scope of Artemis III mission costs.
- *Acquisition management.* NASA's largest, most complex projects, including those that support the Artemis missions, continue to shape the agency's portfolio. When these projects exceed their cost baselines and require cost reserves to meet their funding needs, it has a cascading effect on other projects. NASA officials are exploring ways to better manage this project cost and schedule growth.

January 17, 2024

Chairman Babin, Ranking Member Sorensen, and Members of the Subcommittee:

Thank you for the opportunity to discuss the National Aeronautics and Space Administration's (NASA) efforts to return astronauts to the surface of the moon and, ultimately achieve human exploration of Mars. In the fiscal year 2024 President's budget request, NASA requested at least $38 billion over the next 5 years to support this ambitious undertaking, known collectively as the Artemis missions. The projects supporting the Artemis missions are complex and specialized, and often push the state of the art in space technology. Executing Artemis missions will require extensive coordination across several NASA programs to ensure systems operate together seamlessly and safely. The Artemis missions will also partner with contractors to develop, demonstrate, and produce critical components as part of a strategy to leverage commercial investment and interest in space technology.

We previously highlighted NASA's progress toward achieving the lunar landing mission, such as establishing integration processes and completing some lunar program development activities. We also reported on the challenges NASA faces in developing and integrating these systems and missions.[27] Improving acquisition management—which has been a long-standing challenge at NASA—will play a key role in successfully executing the Artemis enterprise.[28]

You asked us to testify today on our work examining NASA's lunar programs. My statement focuses on (1) progress NASA has made on its Artemis missions, and (2) challenges the agency faces in conducting these missions.

This statement is based on our previously issued reports on NASA's Artemis efforts, including reports that focus on the lunar programs necessary to support the Artemis III missions and our annual assessment of NASA's major projects. We also compared original Artemis mission dates from previously obtained NASA documentation to the new dates announced by NASA in January 2024 to determine any delays. The reports cited throughout this statement include detailed information on their scope and methodology.

We conducted the work on which this statement is based in accordance with generally accepted government auditing standards. Those standards require that we plan and perform the audit to obtain sufficient, appropriate evidence to provide a reasonable basis for our findings and conclusions based on our audit objectives. We believe that the evidence obtained provides a reasonable basis for our findings and conclusions based on our audit objectives.

Background

Key Elements of NASA's Planned Return to the Moon

The goal of NASA's Artemis enterprise is to return U.S. astronauts to the surface of the moon, establish a sustained lunar presence, and, ultimately, achieve human exploration of Mars. To do so, NASA programs are developing multiple highly complex and interdependent systems that will need to be integrated to support individual Artemis missions.

---

[27] GAO, NASA Lunar Programs: Improved Mission Guidance Needed as Artemis Complexity Grows, GAO-22-105323 (Washington, D.C.: Sept. 8, 2022); and NASA Artemis Programs: Crewed Moon Landing Faces Multiple Challenges, GAO-24-106256 (Washington, D.C.: Nov. 30, 2023).

[28] GAO, High-Risk Series: Efforts Made to Achieve Progress Need to be Maintained and Expanded to Fully Address All Areas, GAO-23-106203 (Washington, D.C.: Apr. 20, 2023).

- The Artemis I and II missions are the first uncrewed and crewed demonstration missions, respectively, of the Space Launch System (SLS) launch vehicle, the Orion Multi-Purpose Crew Vehicle (Orion), and the associated ground systems, known as Exploration Ground Systems (EGS).
- The Artemis III mission will leverage contracts with commercial companies to develop the human landing system (HLS) and space suits. NASA awarded firm-fixed-price indefinite delivery, indefinite quantity contracts to two companies, SpaceX and Axiom Space, to develop these capabilities.[29]
  - SpaceX is developing the HLS, which will provide crew access to the lunar surface and demonstrate initial capabilities required for deep space missions.
  - Axiom Space is developing modernized space suits, which consist of a combination of a pressure garment and life support components that together will provide capacity for at least 8 hours of lunar surface activity.
  - The Artemis IV and later missions plan to focus on establishing a sustainable lunar presence. For example, NASA is developing a lunar orbiting outpost—called the Gateway—as a habitat and safe work environment for astronauts. NASA also plans to use the SLS Block 1B and Mobile Launcher 2, which will provide additional capability in these later missions.
  - NASA stated that later missions on and around the moon will help prepare for the types of mission durations and operations it will experience on human missions to Mars.[30]

See figure 1 for the programs needed to accomplish the Artemis missions.

---

[29] NASA also awarded a task order to Collins Aerospace to begin development activities on a space suit capability for Artemis III. However, we did not include these activities in our work upon which this statement is based.

[30] NASA, Artemis Plan: NASA's Lunar Exploration Program Overview (September 2020).

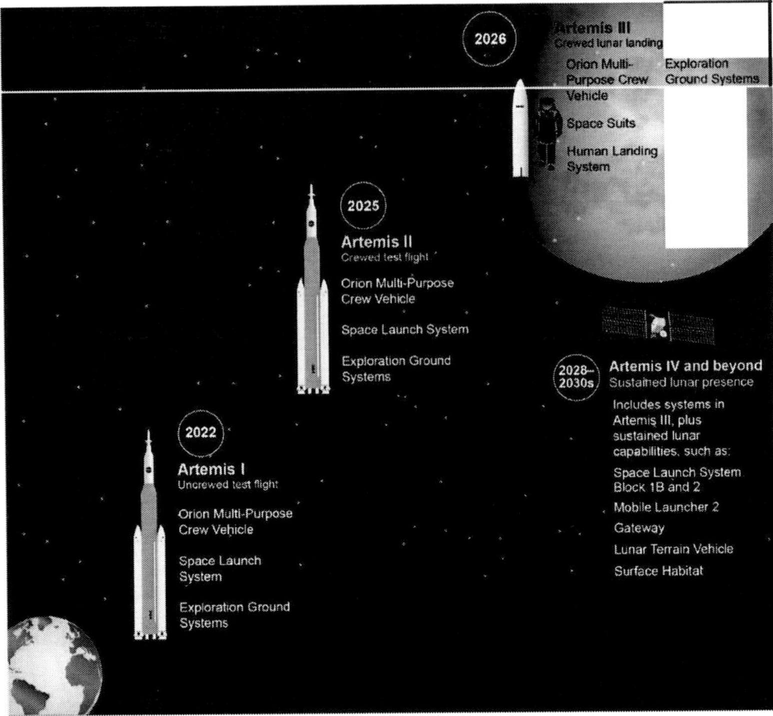

Source: GAO presentation of NASA documentation. I GAO-24-107249.

Figure 1. Key NASA Programs Supporting Artemis Missions.

NASA Public-Private Partnerships

NASA has expanded its effort to contract with commercial companies, especially for its human spaceflight efforts. For example, NASA established the Commercial Crew and Cargo Program Office in 2005 to encourage the growth of the private spaceflight sector in the U.S. According to NASA, the public-private partnerships established by this program office represented a new way of doing business in the realm of human spaceflight.

NASA has continued to build on this experience to support the Artemis missions to return humans to the lunar surface. For example:

- The HLS program is using commercial partnerships to develop and jointly deploy a landing system to transport humans to and from the lunar surface. NASA expects that its commercial partners will heavily leverage NASA technology and expertise throughout the development process, leading to a lunar transportation system that will deliver humans to the lunar surface. NASA also expects that its commercial partners will develop and demonstrate a more sustainable HLS for subsequent crewed missions. In July 2021, NASA exercised a $2.9 billion option on its contract with SpaceX to provide crew access to the lunar surface and demonstrate initial capabilities for deep space missions.[31]

- NASA's Extravehicular Activity (EVA) Development project, which oversees space suit development, is also using commercial partnerships to develop a modernized space suit and associated hardware for lunar surface exploration. In May 2022, NASA awarded firm-fixed-price indefinite delivery, indefinite quantity contracts to Axiom Space and Collins Aerospace. These companies are to provide safe and reliable commercial extra-vehicular activities in microgravity and partial gravity environments on the International Space Station and the lunar surface for Artemis missions.[32] In September 2022, NASA issued a $229 million order under Axiom's contract for the development and demonstration of a suit for lunar surface activities. Axiom is required to provide space suits that will allow crew to successfully perform exploration and science missions on the lunar surface during the Artemis III mission.

---

[31] NASA first awarded the HLS contract to three providers in May 2020. In April 2021, NASA announced the selection of SpaceX for the award of the contract to develop the Artemis III human landing system. After the award, Blue Origin and Dynetics filed bid protests with GAO, which GAO denied in July 2021. GAO, *Blue Origin Federation, LLC; Dynetics, Inc.-A Leidos Company,* B-419783; B-419783.2; B-419783.3; B-419783.4, July 30, 2021, 2021 ¶ CPD 265 (Washington, D.C.: July 30, 2021). Subsequently, in August 2021, Blue Origin filed a complaint with the U.S. Federal Court of Claims. The court dismissed this complaint in November 2021. Blue Origin Fed. LLC v. United States, Fed. Cl., No. 21-1695C (Nov. 4, 2021).

[32] An indefinite delivery, indefinite quantity contract provides for an indefinite quantity, within stated limits, of supplies or services during a fixed period. The government places orders for individual requirements. FAR 16.504(a).

NASA Continues to Make Progress Demonstrating Capabilities Needed for the Lunar Landing Mission

Since we reported on the status of the Artemis missions and programs in March 2022, NASA has demonstrated a number of initial capabilities needed to support the lunar landing mission.[337] Examples of key events include the following:

- Artemis I successfully launched on November 16, 2022, with the Orion capsule safely returning to Earth on December 11, 2022. SLS and EGS demonstrated their initial capabilities during this first test flight.
- Artemis II integration and testing with the Orion crew capsule is ongoing. In October 2023, NASA joined together the Orion crew module and service module. Now that the crew and service modules are integrated, the team will power up the combined crew and service module for the first time. After power on tests are complete, Orion will begin altitude chamber testing, which will put the spacecraft through conditions as close as possible to the environment it will experience in the vacuum of deep space.

  NASA plans to conduct several key integration and test events, for example, ground system testing of the new launch pad systems.

  Teams will conduct a variety of tests and continue ground systems upgrades. These preparations include testing the pad's emergency egress system. After testing at the pad is complete, the mobile launcher will travel to the Vehicle Assembly Building in preparation for rocket stacking operations ahead of launching Artemis II.
- NASA continues to make progress on its integration and risk management plans for the *Artemis III* mission. In September 2022, we found that NASA had established several mechanisms for identifying and tracking Artemis III risks—including a risk database, scorecard, and cross-program risk reviews—and had begun implementing them.[34]

---

[33] GAO, *NASA Lunar Programs: Moon Landing Plans Are Advancing but Challenges Remain*, GAO-22-105533 (Washington, D.C.: Mar. 1, 2022).

[34] GAO-22-105323.

Further, in November 2023, we found that NASA had made progress completing several important milestones with contractors to develop the HLS and space suits needed to support the Artemis III mission.[35]

- NASA and SpaceX completed several important milestones and made progress in designing and testing components of the HLS Starship. SpaceX is currently developing a commercial Starship vehicle to transport humans and cargo to low-Earth orbit, the moon, and Mars. The HLS Starship system consists of the SpaceX Super Heavy booster (launch vehicle) and HLS Starship (the vehicle that provides crew access to the lunar surface). SpaceX is also developing a propellant tanker and on-orbit propellant depot for its lunar landing mission concept.

  Additionally, SpaceX conducted launches of its commercial Starship in April and November 2023. The two test flights provided SpaceX with early in-flight data on the engines, vehicle tanks, and primary structures, among other things. These test flights are important steps towards eventually testing the lander's propellant transfer capabilities in space. We found that these were key development tests for achieving the planned crewed landing.
- Axiom made progress in developing the space suits by completing several milestones, including the mission concept review in December 2022 and the Certification Baseline Review in March 2023. To deliver and demonstrate lunar surface space suits and associated systems, Axiom is leveraging many aspects of NASA's previously developed design. According to Axiom representatives, they entered preliminary design review in September 2023 and completed the crew capability assessment.

NASA Faces Challenges Related to Artemis Schedule, Cost, and Acquisition Management

While NASA continues to develop capabilities needed to support its Artemis efforts, the agency faces several challenges. These include an ambitious Artemis III schedule, a lack of transparency into Artemis mission and program costs, and other acquisition management challenges.

---

[35] GAO-24-106256.

Updated Artemis III Mission Time Frames Acknowledge Remaining Work

In January 2024, NASA announced new mission dates. It shifted Artemis II from the most recent estimate of November 2024 to September 2025 and Artemis III from the most recent estimate of December 2025 to September 2026. NASA officials stated that this shift will allow additional time to complete testing and remaining technical work. The revised NASA estimates show a 2-year delay from the original launch dates for the Artemis II and Artemis III missions. Table 1 depicts the changes to the planned mission dates.

Table 1. Original and planned dates for first three Artemis missions as of January 2024

Artemis mission	Originally planned launch date	Current planned launch date
Artemis I	November 2018	Successfully launched November 2022
Artemis II	April 2023	September 2025
Artemis III	September 2024	September 2026

Source: GAO analysis of NASA documentation. | GAO-24-107249.

In our November 2023 report, we found that a variety of factors made the previous December 2025 lunar landing date unlikely.[36] These factors included an ambitious schedule, delays to key events, and remaining technical work. Specifically, we found that, if the HLS development takes as many months as NASA major projects do on average, the Artemis III mission would likely occur in January 2027. Our analysis found that past NASA projects that have launched since 2010 took 92 months from project start to launch, while NASA's planned development time for the HLS was 79 months. Additionally, we found that the HLS program and SpaceX had delayed eight out of 13 key events by between 6 and 13 months. According to NASA, the updated mission time frames will allow SpaceX and Axiom additional time for testing and refinements ahead of the Artemis III mission.

The HLS and Extravehicular Activity and Human Surface Mobility programs will need to complete a significant amount of complex technical work to achieve the planned September 2026 lunar landing goal. For example, SpaceX has remaining development work on both the Raptor engine and on-orbit propellant transfer technology to mature them.

---

[36] GAO-24-106256.

Likewise, Axiom has significant work to complete, including maturing critical technologies for the space suit life support system, procuring suit components that are susceptible to supply chain delays, and qualifying the suit for flight.

Mission and Program Costs Are Not Transparent

To date, NASA has not yet prepared an estimate of how much the Artemis III mission—or subsequent Artemis missions—are likely to cost. Similarly, it does not plan to measure the production costs for the SLS rockets that constitute a significant proportion of future Artemis-related costs. As such, decisionmakers will have limited information available to help inform decisions on the overall lunar investment.

- Artemis III mission costs. In December 2019, we found that NASA estimated that Artemis III may cost between $20 billion and $30 billion, but the agency did not plan to establish an official cost estimate.[37] At that time, we recommended that it do so. NASA agreed with the recommendation and indicated it would provide a preliminary cost estimate for the Artemis III mission by the end of calendar year 2020. NASA did not do so at that time. Subsequently, in February 2023, NASA officials stated that they are developing a methodology to provide Congress with an assessment of costs for each Artemis mission. NASA officials stated that the mission estimates will include the cost of hardware production, integration costs, and operations costs, but did they not provide a time frame for when this would be completed. Implementing our previous recommendations to develop a life-cycle cost estimate for the Artemis III mission as a whole will enable NASA to effectively monitor total mission costs and give Congress valuable insight into mission affordability when making decisions about each year's budget.
- SLS production costs. In September 2023, we found that NASA does not plan to measure production costs for the SLS program.[38] Since SLS's first launch for Artemis I in November 2022, NASA plans to

[37] GAO, *NASA Lunar Programs: Opportunities Exist to Strengthen Analyses and Plans for Moon Landing*, GAO-20-68 (Washington, D.C.: Dec. 19, 2019).

[38] GAO, *Space Launch System: Cost Transparency Needed to Monitor Program Affordability*, GAO-23-105609 (Washington, D.C.: Sep 7, 2023).

spend billions of dollars to continue producing multiple SLS components—such as core stages and rocket engines—needed for future Artemis missions. These ongoing production costs are not captured in a cost baseline, which limits transparency and efforts to monitor the program's long-term affordability. This is important because the production and other costs for the SLS program account for more than one-third of NASA's budget request for programs required to return to the moon. For example, in the President's budget submission for fiscal year 2024, NASA requested $6.8 billion for the five programs that will be required for Artemis III. The SLS program accounted for about $2.5 billion, or 37 percent of that request. Implementing our prior recommendations to establish cost and schedule baselines that capture these ongoing, recurring production costs could improve transparency into the program.

Acquisition Management Challenges

NASA has made improvements to its acquisition management policies and practices—a long-standing challenge at NASA—in recent years. However, it still faces challenges in its ability to manage its costliest and most complex programs, such as those that are critical to support the Artemis missions. Several of the key improvements we have reported on since March 2022 include the following examples:

- In our June 2022 report, we found that the agency institutionalized some strategic, senior-level reviews to understand and address the ongoing risks that its portfolio may face.[39] For example, the agency holds monthly reviews chaired by the NASA Associate Administrator to discuss issues requiring leadership awareness and identify solutions to challenges as they arise. NASA officials told us that senior management periodically assesses mission directorate portfolios, focusing on Category 1 and other highly visible programs and projects during these meetings.[40]

---

[39] GAO, NASA: Assessments of Major Projects, GAO-22-105212 (Washington, D.C.: June 23, 2022).

[40] Projects designated as Category 1 are NASA's highest priority projects and generally have life-cycle costs over $2 billion.

- In August 2022, NASA updated its corrective action plan as part of its efforts to address recent programmatic performance and its inclusion in our biennial High-Risk Report.[41] The plan describes a number of actions the agency intends to take to improve acquisition and program management.
- In our April 2023 High-Risk Update, we found that NASA completed several initiatives to strengthen its cost and schedule estimating capacity and is embracing tools to support better management practices.[42]
- We noted, however, that NASA will need to identify ways to improve its management of Category 1 projects to continue reducing acquisition risk and demonstrating progress.

These Category 1 projects drive cumulative cost performance for the entire portfolio when they overrun their baselines. In our 2023 assessment of major NASA projects, we found that NASA anticipated setting baselines for six Artemis programs.[43] As these projects enter the portfolio, they will drive the agency's acquisition performance over the next several years. NASA senior leaders said that recent efforts intended to help control project cost and schedule growth include having projects (1) document when they deviate from the agency's policy for establishing cost and schedule baselines and (2) develop plans to remove work if cost growth or schedule delays occur. These officials said they plan to explore additional ways to control project costs and schedules, specifically for Category 1 projects.

In summary, NASA has made important progress on its Artemis efforts, but completing the lunar landing mission remains challenging. NASA needs to continue to find ways to better manage the cost of its most complex programs. Further, NASA has not yet determined how much Artemis III and future Artemis missions are likely to cost, limiting critical information needed by decision-makers about the lunar mission.

Implementing our past recommendations will help NASA to improve in these critical areas.

---

[41] GAO-23-106203.

[42] GAO-23-106203.

[43] GAO, *NASA: Assessments of Major Projects*, GAO-23-106021 (Washington, D.C.: May 31, 2023).

Chairman Babin, Ranking Member Sorensen, and Members of the Subcommittee, this completes my prepared statement. I would be pleased to respond to any questions that you may have at this time.

GAO Contact and Staff Acknowledgments

If you or your staff have any questions about this testimony, please contact William Russell, Director, CNSA at (202) 512-4841or russellw@gao.gov. Contact points for our Offices of Congressional Relations and Public Affairs may be found on the last page of this statement.

GAO staff who made key contributions to this testimony are Kristin Van Wychen (Assistant Director); Erin Roosa (Analyst-in-Charge); John Armstrong; Breanne Cave; Edward Harmon; Tonya Humiston; Erin Kennedy; John Ortiz; Sylvia Schatz; Kate Sharkey; Juli Steinhouse; Kevin Walsh; Alyssa Weir; and Tonya Woodbury.

GAO's Mission

The Government Accountability Office, the audit, evaluation, and investigative arm of Congress, exists to support Congress in meeting its constitutional responsibilities and to help improve the performance and accountability of the federal government for the American people. GAO examines the use of public funds; evaluates federal programs and policies; and provides analyses, recommendations, and other assistance to help Congress make informed oversight, policy, and funding decisions. GAO's commitment to good government is reflected in its core values of accountability, integrity, and reliability.

Connect with GAO

Connect with GAO on Facebook, Flickr, Twitter, and YouTube. Subscribe to our RSS Feeds or Email Updates. Listen to our Podcasts. Visit GAO on the web at https://www.gao.gov.

To Report Fraud, Waste, and Abuse in Federal Programs

Contact FraudNet:

Website: https://www.gao.gov/about/what-gao-does/fraudnet Automated answering system: (800) 424-5454 or (202) 512-7700.

Congressional Relations

A. Nicole Clowers, Managing Director, ClowersA@gao.gov, (202) 512-4400, U.S. Government Accountability Office, 441 G Street NW, Room 7125, Washington, DC 20548.

Public Affairs

Chuck Young, Managing Director, youngc1@gao.gov, (202) 512-4800.

U.S. Government Accountability Office, 441 G Street NW, Room 7149 Washington, DC 20548.

Strategic Planning and External Liaison

Stephen J. Sanford, Managing Director, spel@gao.gov, (202) 512-4707.

U.S. Government Accountability Office, 441 G Street NW, Room 7814, Washington, DC 20548.

William Russell

William Russell is a Director in GAO's Contracting and National Security Acquisitions teama. He oversees a portfolio of issues related to NASA acquisitions, DOD supply chains, and federal contracting.

William joined GAO in 2002 and has served as Acting Director in GAO's Homeland Security and Justice team, managing a portfolio of issues related to aviation security, surface transportation security, and DHS research and development.

William earned a master's degree in foreign affairs from the University of Virginia and a bachelor's degree in political science from Virginia Commonwealth University.

Testimony before the House of Representatives Subcommittee on Space and Aeronautics, Committee on Science, Space, and Technology

Key Challenges Facing NASA's Artemis Campaign Statement of George A. Scott Acting Inspector General National Aeronautics and Space Administration

Chairman Babin, Ranking Member Sorensen, and Members of the Subcommittee:

Our mission is to provide independent, objective, and comprehensive oversight of NASA's programs and projects to help ensure that the Agency operates with transparency, efficiency, and accountability. As part of this mission, we provide oversight on significant challenges facing NASA and impacting the Artemis campaign.

After more than a decade of preparation and several delays, in December 2022 NASA successfully completed Artemis I—an uncrewed test flight to lunar orbit. Artemis I was a significant achievement for NASA, providing important data and lessons learned from the testing of hardware, software, processes, and teams that will help prepare NASA for future Artemis missions. Despite this achievement, our oversight has identified several interrelated challenges NASA must address to achieve its ambitious Artemis goals. Of utmost importance is the resolution of technical challenges that could threaten astronaut safety while also addressing historical challenges related to unsustainable costs and a lack of transparency into funding needs.

First, the Artemis campaign's technical challenges. The Agency's immediate challenge is preparing for Artemis II—the first crewed test flight of the Space Launch System (SLS) heavy-lift rocket and Orion Multi-Purpose Crew Vehicle (Orion) system—which will return humans to lunar orbit for the first time in more than 50 years. The Agency continues to analyze mission data from Artemis I and must address a variety of technical challenges to safely fly four astronauts to lunar orbit on their planned 10-day Artemis II mission.

While considered a near-perfect flight by NASA officials, Artemis I revealed technical issues such as the unexpected erosion of protective material on the Orion heat shield. In addition, the Mobile Launcher 1 platform—the ground structure used to assemble, process, transport, and launch the SLS for Artemis I through III—sustained more damage than expected. Just last week, NASA delayed the Artemis II mission to September 2025.

Looking ahead to Artemis III—the mission that will return humans to the surface of the Moon—NASA's commercial partner SpaceX must conduct multiple flight tests and launches of its Human Landing System (HLS) Starship before using its lander variant with astronauts onboard. The HLS requires SpaceX to launch a series of Starship vehicles to establish a "fuel depot" in low Earth orbit to refuel each Starship heading to the Moon. Moreover, under its contract with NASA the company is required to send an uncrewed Starship to the lunar surface and back prior to Artemis III to demonstrate its readiness for a crewed mission. At the same time, NASA must develop additional capabilities including next-generation spacesuits. With last week's announcement, NASA also delayed Artemis III to September 2026 in part to provide additional time to develop SpaceX's HLS Starship and next-generation spacesuits.

For missions beyond Artemis III, the second mobile launcher (ML-2) is a critical part of the infrastructure needed to launch the upgraded SLS Block 1B and Block 2. In June 2022, we reported that the ML-2 project is significantly behind schedule and over budget, jeopardizing launch schedules for Artemis IV and beyond. While the ML-2's first steel components were delivered to Kennedy Space Center in May 2023, we estimate completion of the launcher will not occur until late 2026 at the earliest, 2.5 years behind the project's originally scheduled date.

The second challenge is the Artemis campaign's enormous expense. Overall, we project NASA's total Artemis campaign costs to reach $93 billion between fiscal years 2012 and 2025. We also project the SLS/Orion system and related ground launch infrastructure will cost at least $4.2 billion per launch for the first four Artemis missions, a figure that does not include $42 billion in formulation and development costs spent over the past dozen years to bring these systems to the launch pad.

Development of the systems required to transport humans to the Moon and Mars safely has proven to be especially challenging due to increased costs stemming from significant technical issues, changing requirements, and overly optimistic schedules.

Given these estimated costs and the significant challenge they pose to the long-term sustainability of the Artemis campaign, it is critical that the Agency identify and implement effective ways to reduce costs. This will be especially important as Congress urges NASA to increase the SLS/Orion launch cadence at the same time NASA—and much of the federal government—may be operating under a flat annual budget. Our recent work has shown that some key cost reduction efforts may fall short of expectations.

For example, in May 2023 we reported that NASA is projecting manufacturing cost savings of 30 percent per engine for the SLS starting with production of the seventh of 24 new RS-25 engines. However, these projected savings do not capture overhead and other costs associated with restarting production of the engine, which we estimated to reach $2.3 billion. Likewise, in October 2023 we reported on NASA's efforts to reduce the cost of lunar missions beyond Artemis IV by transitioning management of multiple contractors for production of SLS systems and hardware, as well as systems integration and launch services, to a single contractor service. We found this approach would likely not achieve its cost reduction goals due to a variety of unrealistic assumptions, such as finding customers outside of NASA to use the SLS. Additionally, NASA aims to make its Moon to Mars plan more sustainable by sharing costs with its international partners. However, we found NASA's cost-sharing strategies with its international partners are still evolving and the Agency lacks an overall architecture, or blueprint, that includes cost estimates and responsibilities for international partners beyond Artemis IV.

The final challenge we highlight today is the Artemis campaign's lack of cost and schedule transparency. In particular, NASA still lacks a comprehensive and accurate estimate that accounts for all Artemis costs. For example, we previously reported that NASA had neither established life-cycle costs nor made cost and schedule commitments for some of the programs supporting the Artemis campaign. By failing to do so, the Agency is circumventing congressional requirements for reporting and tracking such costs. We continue to believe the Agency needs to provide full visibility into its investments as it begins a multi- decade Moon to Mars initiative at a cost that could easily reach into the hundreds of billions of dollars. As the programs that support these exploration missions transition from development to production and operation, it is critical that NASA establish credible, complete, and transparent cost and schedule estimates from which they can measure success and be accountable to Congress and other stakeholders.

Over the past two years, the OIG has issued nine audit reports that examine issues critical to NASA's effort to land humans on the Moon as a prelude to a crewed Mars mission. We assessed NASA's transition of the SLS to a commercial services contract, the Artemis supply chain, communication infrastructure, SLS engine and booster contracts, partnerships with international space agencies, ground systems and launch infrastructure, cost estimating and reporting practices, management of the Agency's astronaut corps, and management of the Artemis missions. Below, we summarize these reports, findings, and recommendations.

NASA's Transition of the Space Launch System to a Commercial Services Contract (IG-24-001, October 2023)

In an effort to increase the affordability of the Artemis campaign, NASA is preparing to award a sole- sourced services contract, known as the Exploration Production and Operations Contract (EPOC), to Deep Space Transport, LLC (DST)—a newly formed joint venture of The Boeing Company and Northrop Grumman Systems Corporation—for the production, systems integration, and launch of at least 5 and up to 10 SLS flights beginning with Artemis V scheduled for 2029.

We found that despite NASA's noteworthy adjustments to the EPOC transition plan and its affordability initiatives, the price of the SLS Block 1B rockets will not be significantly reduced through a sole-source contract with DST. NASA's aspirational goal is to achieve a 50 percent cost savings over current SLS production costs using DST, which by our calculation would reduce the contract cost of a single SLS rocket from $2.5 billion to $1.25 billion. Our analysis shows this goal cannot be achieved and the production cost alone will remain over $2 billion. We reach this conclusion after examining what we believe are unrealistic assumptions on NASA's part. First, the Agency expects to achieve cost savings through reduced SLS production costs under a contract with DST. However, ongoing affordability efforts by SLS contractors to reduce the workforce and improve manufacturing processes have yet to achieve cost savings on the high-cost stages and RS-25 engine contracts. Second, DST expects to drive down costs by increasing the SLS production rate and building more SLSs for non-NASA customers such as the Department of Defense and commercial entities. However, thus far other potential users have declined to use the SLS due to lower-cost alternatives. Finally, NASA's ability to negotiate less costly services with

DST will be hindered by the lack of competition given EPOC is sole sourced to the existing SLS contractors.

The OIG made seven recommendations to improve the sustainability of the SLS system.

NASA's Management of the Artemis Supply Chain (IG-24-003, October 2023)

Each of NASA's Artemis-related programs rely on specialized parts supplied by contractors and subcontractors from across the United States and around the world. NASA's contractors employ a network of subcontractors and suppliers to provide the hardware, raw materials, electronic parts, and other resources needed to fulfill their contracts. To support the Artemis campaign, NASA obligated approximately $40 billion to 860 contractors from fiscal years 2012 to 2022.

We found that NASA and its prime contractors continue to experience challenges obtaining key components and necessary supplies to meet Artemis goals resulting in cost increases and schedule delays. Supply chain delays and disruptions over the past several years have resulted from a variety of factors outside the Agency's control, from the COVID-19 pandemic to inflation of wages and material costs to the Russia-Ukraine conflict. That said, several factors related to managing Artemis supply chain issues are within NASA's purview. Most importantly, this includes the Agency's lack of visibility into its critical suppliers, with many Artemis programs and projects not tracking their prime contractors' supply chain impacts. Even when issues with subcontractors and suppliers are identified, performance challenges are not shared across Artemis programs to enable effective procurement decisions. To its credit, NASA is undertaking efforts to better understand supply chain issues and manage them more proactively, but these initiatives are still in the early stages.

The OIG made seven recommendations to improve NASA's management and visibility into its supply chain.

Audit of NASA's Deep Space Network (IG-23-016, July 2023)

NASA relies on its Deep Space Network (DSN) to provide communication links that guide and control spacecraft such as the Orion and bring back images and other data from missions such as the James Webb Space Telescope. The DSN consists of three communications facilities in the United States, Spain, and Australia that use antennas to communicate with spacecraft located between 10,000 miles from Earth to beyond the edge of the solar system.

We found DSN antennas are operating at capacity and are oversubscribed—meaning more time is requested by missions than the network's current capacity can provide—with demand exceeding supply at times by as much as 40 percent. The Agency's crewed Artemis missions to the Moon will require increasingly higher amounts of bandwidth and further constrict the network's ability to meet growing mission demands. As NASA pivots toward extended human exploration of the Moon, the Agency may need to give DSN capacity to priority missions in critical phases, such as launches, while other missions make do with limited or no data during those periods. NASA's primary solution to address the DSN's capacity issues is to construct additional antennas and make upgrades to existing infrastructure.

However, these efforts are behind schedule and over budget, experiencing nearly 5 years of delays, only partial completion of two phases of construction, and an expected 68 percent cost increase.

The OIG made four recommendations to ensure NASA's progress towards upgrading the Agency's DSN and the network's ability to support current and future mission requirements.

NASA's Management of the Space Launch System Booster and Engine Contracts (IG-23-015, May 2023)

Key to NASA's Artemis campaign is development of the SLS—a two-stage, heavy-lift rocket with two boosters and four RS-25 engines that will launch the Orion into space. From fiscal years 2012 through 2025, NASA's overall Artemis investment is projected to reach $93 billion, of which the SLS Program costs represent $23.8 billion spent through 2022. This audit examined two SLS booster contracts with Northrop Grumman and two RS-25 engine contracts with Aerojet Rocketdyne. We found that NASA is experiencing significant scope growth on both contracts, as well as approximately $6 billion in cost increases and over 6 years in schedule delays. As a result of the cost and schedule increases under these four contracts, we calculate NASA will spend $13.1 billion through 2031 on boosters and engines.

We found long-standing management issues—including underestimating the scope and complexity of work, concurrent development and production activities, inadequate procurement workforce, and inappropriate use of award fees—caused the cost increases and schedule delays. Further, NASA's poor contract management practices are impacting the SLS Program and Artemis campaign, causing us to question $49.9 million in costs and award fees. Facing continuing cost and schedule increases, we found NASA is undertaking efforts to make the SLS more affordable. Under the RS-25 Restart and Production contract, NASA and Aerojet Rocketdyne are projecting manufacturing cost savings of 30 percent per engine starting with production of the seventh of 24 new engines. However, those savings do not capture overhead and other costs, which we currently estimate at $2.3 billion. For SLS boosters, NASA is procuring 10 boosters on a fixed-price-incentive-fee basis starting with Artemis IV—an important step in its affordability initiatives—but any additional requirements will limit these projected cost savings.

The OIG made eight recommendations to help increase transparency, accountability, and oversight of the SLS booster and engine contracts and NASA's affordability efforts.

NASA's Partnerships with International Space Agencies for the Artemis Campaign (IG-23-004, January 2023)

NASA's partnerships with international space agencies are critical to achieving a robust and sustainable presence on the Moon as a precursor to a human mission to Mars. Key early Artemis commitments from partner agencies include the provision of a Gateway habitat, communications satellites, spacecraft service modules, external robotics, astronauts, and lunar rovers.

We found, however, future international cooperation for Artemis may be hindered by a variety of factors. This includes the Agency's lack of an overarching strategy to coordinate Artemis contributions from international space agencies and entities. While the architecture, or blueprint, for the first three Artemis missions is well established, an overall architecture beyond Artemis IV for lunar exploration that includes estimated costs to be borne and responsibilities assumed by its international partners is not yet established. As a result, partners have insufficient information to work with their governments to develop their own budgets and identify potential contributions to the Artemis effort. In addition, U.S. export control regulations can be overly complex and restrictive which may limit NASA's international collaborations on Artemis. Finally, NASA's cost sharing strategies with its international partners for Artemis are still evolving and, in contrast to International Space Station operations where international partners contribute almost 25 percent of the costs, we estimate that less than 6 percent of the human space flight mission costs will be borne by international partners for the first three Artemis missions.

The OIG made ten recommendations to increase the effectiveness and affordability of Artemis integration efforts with international partners.

NASA's Management of the Mobile Launcher 2 Contract (IG-22-012, June 2022)

Key to NASA's goals of sustaining a human presence on the Moon and future exploration of Mars is the Agency's development of two mobile launchers that will serve as the ground structure to assemble, process, transport to the pad, and launch various iterations of the integrated SLS/Orion system into space. In 2019, NASA awarded a cost-plus contract to Bechtel National, Inc. to design, build, test, and commission a second mobile launcher to support larger variants of the SLS beginning with the Artemis IV mission. Valued at $383 million, the original contract had a performance period from July 2019 through March 2023.

We found that for completion of contract requirements and delivery of an operational ML-2, Bechtel estimated it would need an additional $577.1 million, for a total cost of $960.1 million, and an October 2025 delivery date rather than March 2023 as initially planned. Additionally, we found ML-2's substantial cost increases and schedule delays could be attributed primarily to Bechtel's poor performance on the contract, with more than 70 percent ($421.1 million) of the contract's cost increases and over 1.5 years of the delays experienced being related to the company's performance. These increases and delays were further compounded by NASA's management practices and decision to award the contract before the SLS's Exploration Upper Stage requirements were finalized. Further, NASA's usage of award fees did not improve Bechtel's performance.

The OIG made five recommendations to improve management of the ML-2 contract and contractor performance. In September 2023, the OIG initiated a new audit to examine the actions NASA is taking to control future cost growth and schedule delays.

NASA's Cost Estimating and Reporting Practices for Multi- Mission Programs (IG-22-011, April 2022)

NASA has a long history of groundbreaking accomplishments but has struggled to establish credible cost estimates for some major acquisitions; particularly, human space flight missions, which are comprised of multiple programs with numerous deliverables—like rockets and spacecraft—stretching over many years. As a result, Congress and other stakeholders lack meaningful visibility into the complete costs of NASA's major acquisitions. Without adequate transparency, it is difficult for stakeholders to hold the Agency accountable for these large, years-long expenditures of taxpayer funds.

We found that Congress is not receiving the federally mandated cost and schedule information it needs to make fully informed funding decisions for NASA's programs—specifically, the SLS, Orion, and Exploration Ground Systems—that support Artemis. NASA only made cost and schedule commitments to Congress to demonstrate the initial capability of each system. Even though NASA has multiple Artemis missions planned, it has not adjusted the three programs' life-cycle cost estimates or commitments to account for future missions. The result is incomplete cost estimates and commitments for these programs and missions.

In August 2021, the Agency made an update to NASA Procedural Requirements 7120.5F, NASA Space Flight Program and Project Management Requirements, which establishes the requirements, life-cycle processes, and procedures NASA uses to formulate and implement space flight programs and projects. NASA stated that it intended to establish new policies and procedures that would provide additional transparency for major programs with multiple deliverables and unspecified end points. Instead, it codified its poor cost estimating and reporting practices in a new policy that fails to comply with Title 51 of the United States Code, National and Commercial Space Programs, which requires the Agency to annually provide an estimate of the life-cycle cost for major programs, with a detailed breakout of the development cost, program reserves, and an estimate of the annual costs until development is completed. The policy also weakens NASA's ability to account for some risks in programs consisting of multiple projects, potentially affecting cost and schedule if risks are unidentified in the estimates.

The OIG made seven recommendations to ensure that all major programs and activities are reported to Congress in accordance with Title 51. In July

2023, NASA informed the OIG that it would not implement four of the recommendations.

NASA's Management of Its Astronaut Corps (IG-22-007, January 2022)

As NASA enters a new era of human space flight, effective management of its astronaut corps is critical to the Agency's success. With the upcoming crewed Artemis II mission, the margin of time available to identify skillset needs, recruit and hire additional astronaut candidates, develop a framework for Artemis training, and adjust current processes for sizing, aligning, training, and assigning its astronaut corps is quickly diminishing.

We found NASA's processes used to size, train, and assign the astronaut corps are primarily calibrated toward meeting the current needs of the International Space Station. However, work has begun to align NASA's astronaut corps to Artemis mission needs. As the Agency prepares for crewed Artemis missions, astronaut training needs will change. While the Astronaut Office is in the process of developing a framework for Artemis training, it has not been formally chartered. Delays in moving beyond the current Space Station-focused approach increases the risk of delays in developing the necessary training to meet Artemis mission goals.

Additionally, as NASA moves to deep space Artemis missions, it has begun to review its policies and conduct additional studies on human health impacts from longer duration missions and missions beyond low Earth orbit. If the nature of Artemis missions medically disqualifies certain astronauts as a result of exceeding the Agency-set maximum level of radiation exposure because of the duration of the mission beyond low Earth orbit, NASA may need to adjust its astronaut corps size and assignment process.

The OIG made four recommendations to help ensure the astronaut corps is aligned to meet current and future mission needs.

NASA's Management of the Artemis Missions (IG-22-003, November 2021)

We found that NASA is projected to spend $93 billion on the Artemis campaign from fiscal years 2012 through 2025. However, as a result of

NASA's decision not to classify Artemis as a formal program under the Agency's Space Flight Program and Project Management Requirements, Artemis officials were not required to develop an official Artemis-wide full life-cycle cost estimate. By failing to develop an official cost estimate that includes all relevant costs, NASA lacks transparency of the true funding requirements for a long-term Artemis effort.

Multiple factors contribute to the high cost of exploration system development programs, including the use of sole-source, cost-plus contracts; the inability to definitize key contract terms in a timely manner; and the fact that except for the Orion capsule, its subsystems, and supporting launch facilities, all components are expendable and "single use" unlike emerging commercial space flight systems.

For HLS, NASA has modified its traditional acquisition approach for large space flight programs to reduce costs, encourage innovation, and meet an aggressive schedule for its Artemis lunar landings. HLS will use less standardized milestone reviews and instead utilize other project management techniques throughout development and testing. While the HLS Program leveraged lessons learned and is modeled, in part, after the Commercial Crew Program, HLS tailored its programmatic milestone approach to better fit a services model approach versus the traditional hardware development program. Although these modified approaches have the potential benefit of decreasing costs and encouraging innovation, they also raise the possibility of technical changes later in development plus schedule and performance risks on NASA's human-rated systems.

The OIG made nine recommendations to increase transparency of costs and improve program management.

Conclusion

While Artemis I was a significant achievement for NASA, the Agency faces higher stakes as it flies astronauts on its Artemis II mission. We urge NASA leadership to continue balancing the achievement of its mission objectives and schedule with prioritizing the safety of its astronauts and to take the time needed to minimize any undue risk on this first crewed Artemis mission.

Moving forward, the Agency must continue to look for ways to reduce the enormous costs of the systems required to transport humans to the Moon and Mars safely within the funding allocated by Congress. Failure to reduce these costs will ultimately make the Artemis campaign unsustainable. At the same time, improved transparency of Artemis costs will be crucial to its success. Without NASA fully accounting for and accurately reporting the overall cost of current and future missions, it will be difficult for Congress, the Office of Management and Budget, and the American public to make informed decisions about NASA's long-term funding needs—a key to making Artemis a sustainable venture.

We look forward to helping NASA achieve its ambitious Artemis goals. To that end, we plan to continue examining key challenges in NASA's human exploration missions to the Moon and Mars.

Moving forward, the Agency must continue to look for ways to reduce the enormous costs of the systems required to transport humans to the Moon and Mars safely within the funding allocated by Congress. Failure to reduce these costs will ultimately make the Artemis campaign unsustainable. At the same time, improved transparency of Artemis costs will be crucial to its success. Without NASA fully accounting for and accurately reporting the overall cost of current and future missions, it will be difficult for Congress, the Office of Management and Budget, and the American public to make informed decisions about NASA's long-term funding needs—a key to making Artemis a sustainable venture.

We look forward to helping NASA achieve its ambitious Artemis goals. To that end, we plan to continue examining key challenges in NASA's human exploration missions to the Moon and Mars.

George A. Scott

George A. Scott assumed the role of NASA's Acting Inspector General in January 2024, following five years of service as the Deputy Inspector General.

He most recently held the position of Interim Executive Director of the Pandemic Response Accountability Committee, overseeing over $5 trillion in coronavirus-related pandemic relief funds. Prior to joining NASA, Scott accumulated over 30 years of experience at the Government Accountability Office (GAO). During his time there, he served as the Managing Director of GAO's Homeland Security and Justice Team, where he oversaw GAO's efforts in the Department of Homeland Security (OHS) and the Department of Justice. This included initiatives to secure the nation's borders and infrastructure, as well as strengthen emergency preparedness and response capabilities. Scott has extensive oversight experience in OHS management, acquisition, financial management, and human capital issues. He also led oversight efforts on various justice programs and grants.

Scott, a graduate of the University of North Carolina at Chapel Hill, is a Certified Fraud Examiner. He is an accomplished speaker, having testified numerous times before Congress and presented at various conferences and training events.

Witness Statement for the Hearing Returning to the Moon: Keeping Artemis on Track U.S. House of Representatives Commitee on Science, Space and Technology Space and Aeronautics Subcommitee Rayburn House Office Building Room 2318 Michael D. Griffin, 17 January 2024

NASA, as well as the nation on behalf of which it executes our civil space program, should modify the strategy, tactics, acquisition approach and programmatic structure of human lunar return as it is presently planned. To the topic of this hearing, the Artemis Program should not be "kept on track"; it should be fixed and then prosecuted with all deliberate speed.

Strategic issues first. The agency has awarded fixed-price contracts to SpaceX and Blue Origin to carry out lunar landings for, respectively, $2.9 and $3.4 billion dollars (htps://www.nytimes.com/2021/04/16/science/spacex-moon-nasa.html, htps://www.reuters.com/technology/space/nasa-name-second-company-build-astronaut- lunar-lander-2023-05-19/). The cost of the Apollo Program over the 14-year period from 1960- 73 is estimated to have been $257 B in 2020 U.S. dollars (C. Dreier, An Improved Cost Analysis of the Apollo Program, Space Policy, htps://doi.org/10.1016/j.spacepol.2022.101476). It is reasonable to believe that with the flight experience and space industrial infrastructure that exist today, human lunar missions could and should be executed for considerably less than Apollo. It is grossly unrealistic to suggest that they could be done for 1.5% of Apollo's cost. The award of these unrealistically low fixed-price contracts makes it clear that cost reasonableness was not a factor in ranking these contract awards. The further implication is that the United States is not yet serious about a program that should be regarded as a core national interest – returning U.S. and international partner astronauts to the Moon before our self-declared adversaries can do so.

As in the 1960s, we are again faced with near-term peer competition in space, this time with the Chinese Communist Party and, once again, potentially Russia: (htps://www.newsweek.com/russia-approves-plan-establish-lunar-base-china-1848731). For the U.S. not to be able to put its own and partner astronauts on the Moon, to be watching on the internet while adversary powers do so, makes a statement about a shift of global power and preeminence that we ought not to allow. People and ntions align themselves with leaders; for most of the last 80 years that has been the United States, in partnership with our European and Western Pacific allies. Are we prepared to relinquish that leadership to China? If not, and if we view preeminence in space as part of that leadership and therefore an element of ntional security, then it is again necessary to prioritize urgency of execution.

Underlying the above is a key theme: we cannot separate civil space exploration from national security space. It's one national program, artificially separated at birth by President Eisenhower to demonstrate to the world, and especially to the Soviet Union, that we were a peaceful nation, exploring and developing space for peaceful purposes. But the reality is that the creation of NASA was a national security initiative from the start, a response to the Soviet Union's launch of Sputnik.

National security takes many forms beyond raw power projection. In exploring and developing the space domain we are pioneering the human frontier. Even a casual reading of history shows that every great nation was on the frontiers of its time; this is almost a defining characteristic of great powers. To quote from President Kennedy's "man, Moon, decade" speech, where mankind goes, free men must fully share. The point is that value systems mater. The United States mounted one of the most powerful yet non-aggressive responses in history to the Soviet Union's launch of Sputnik. Had the first satellite been launched by the United Kingdom, the United States might have been a bit chagrined that we weren't first, but the response would simply not have been the same. The values of the United Kingdom and our own are highly aligned; the values of the United States and the Soviet Union were about as antithetical as it was possible to be. This difference was critical to our response to Sputnik, as it should have been.

The reality is that decisions are made, standards are set and values are established on a frontier by the people who show up, not by those who stay home and watch. The society that sets those standards (as we have done for global air transportation since the end of World War II) and establishes the key infrastructure emerges as first among equals, the proper goal for the United States.

Finally, when a society can do things that others cannot it commands a degree of respect that is by itself a valuable national security asset, possibly more so than in many instances of the exercise of "hard power". Quite simply, the very best people want to come to the place where the very best things are being done. It is quite instrucve to observe how many key figures in the Manhatan and Apollo programs were immigrants, a number that was hugely out of proportion to the rest of the population. To quote an observation by former Deputy Under Secretary of Defense for Research and Engineering Dr. Lisa Porter, the United States is a country where a six-sigma individual can flourish. They are the people who create, in the words of another quote atributed to JFK, the rising tide that lifs all boats. Space exploration atracts such people. That is something in which we should take pride and is an asset to be nourished.

These are the forms of national security that NASA enables, and that we should take to heart incrating our national space exploration strategy.

Tactically, the selected mission architectures pose significant concerns. SpaceX's approach requires an impractically large number of orbital refueling operations for even a single lunar mission (Space News, 17 Nov 2023; htps://spacenews.com/starship-lunar-lander-missions-to-require-nearly-20-launches-nasa-says/), while Blue Origin's mission design depends on the development of one of the most difficult enabling technologies for long-duration space flight, zero-boiloff cryogenic fuel storage (htps://en.m.wikipedia.org/wiki/Blue_Moon_(spacecraft)). These architectures feature concepts – cryogenic propellant storage, likely in large depots with low, controllable boiloff – that are critical to long-term, sustainable human space exploration. But while important, their development is unlikely to be completed easily or quickly, and over the last half-century we have used up the time that could have been devoted to the evolution of Apollo-era systems to a more sustainable architecture. Like it or not, we are engaged in a competition with others who do not wish us well; timeliness maters.

There are other concerns as well.

Crew Safety

The present Artemis mission architecture requires staging operations at a Gateway based in a lunar polar near-rectilinear halo orbit (NRHO) with a 6.5-day period and dimensions of 3,000 km x 70,000 km altitude above the lunar surface. This approach is said to offer two significant advantages: the Orion spacecraft, which as discussed below has limited ΔV capability, can get into and back out of this orbit on the way to and from the Moon, and any point on the lunar surface can be accessed from the staging area. The first of these issues can be addressed by far simpler means, discussed below, and the second is not unique to NRHO – it is a characteristic of any polar orbit.

However, these points are trivial in comparison to the major disadvantage of staging from NRHO, which is that immediate return to the Gateway from the lunar surface is possible only on 6.5-day centers. If a lunar crew encounters a problem on the surface that mandates a return to the comparative safety of the Gateway, then depending upon when that problem occurs, a multi-day wait may be required. It is possible in some scenarios to wait in low lunar orbit (LLO), but access to the Gateway is only possible at periodic intervals.

With present technology, flying in space is just barely possible; even in Earth orbit it is both difficult and dangerous. Expeditions to the Moon will be even more demanding. From a safety perspective, no early human lunar mission should knowingly accept the risk of stranding a crew, whether on the surface or in lunar orbit, for days at a time. No mission architecture should be contemplated without, as in Apollo, the capability to leave the surface and rendezvous with a safer habitat within a few hours. Somewhat like the first experience of "wintering over" in Antarctica, when enough lunar surface infrastructure has been emplaced to allow a viable long- term shelter-in-place option to be implemented, the crew abort strategy can be reconsidered. Such is not the case for early human lunar return. The Artemis program has not been designed with this consideration in mind.

Reliability and Mission Risk

Leaving safety aside, mission architectures requiring multiple complex operations in series, such as propellant supply launches and cryogenic fuel transfer, are inherently less reliable than those requiring fewer. The table below makes this point; the left side of the table specifies a postulated reliability for each launch and propellant transfer operation, while across the top is shown a varying number of such operations.

Reliability of One Operation	Number of Operations			
	5	10	15	20
99%	0.95	0.90	0.86	0.82
98%	0.90	0.82	0.74	0.67
97%	0.86	0.74	0.63	0.54
96%	0.82	0.67	0.54	0.44
95%	0.77	0.60	0.46	0.36

The results speak for themselves. Even if (for example) it is assumed that each single operation, e.g., launch plus propellant transfer, can be performed successfully 98% of the time, i.e., with a 1-in-50 failure rate, a mission requiring ten such operations in a specified campaign window will fail to be completed within that window 18% of the time. As a practical mater, mission architectures requiring multiple launch and propellant transfer operations will be very difficult to complete with a reasonable likelihood of overall success. Congress should question whether this is a gamble that, from either the fiscal or national prestige perspective, it wishes to support.

A Lower-Risk Approach: A Two-Launch Solution for Human Lunar Landing

Early lunar return missions that meet NASA's basic requirements – four people on the surface for a week at any location – can be achieved using technology and systems that are largely available today. One straightforward approach is discussed below. It requires two SLS Block 2 heavy lift launches, each carrying a Centaur III upper stage; an Orion command and service module; and a two-stage storable-propellant lunar lander, yet to be designed. A schematic view of this approach is shown below:

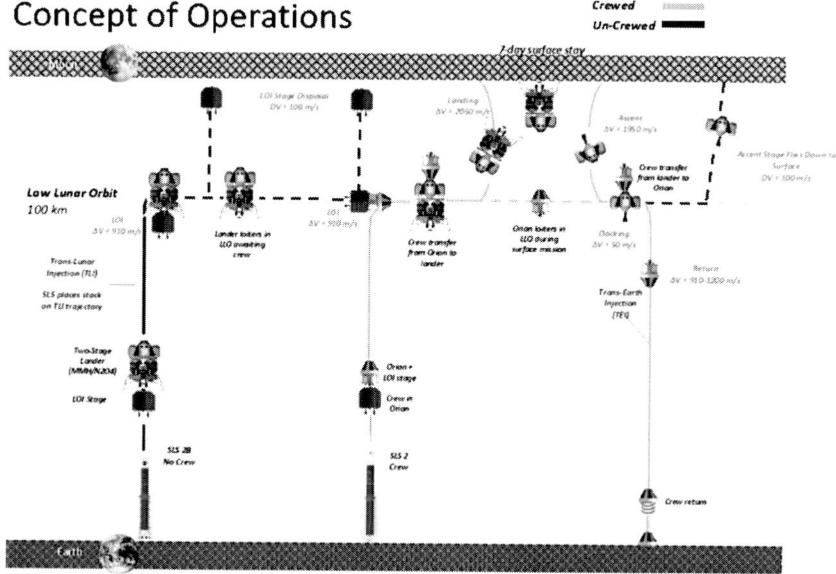

Mission Sequence

1) A payload stack consisting of a partially fueled Centaur III upper stage and the fully fueled but uncrewed Lander is launched as cargo on the SLS Block 2B cargo variant with the capability to put about 45 metric tons (mT) into a trans-lunar insertion (TLI) trajectory.
2) The Centaur III is fueled with sufficient propellant (including allowance for boiloff) to provide a ΔV of about 1 km/s for the payload stack and is used as a lunar orbit insertion (LOI) stage to deliver the Lander to LLO to await the crew.
3) At a later time, the crew is launched on an SLS Block 2 crew variant (41 mT to TLI) to LLO in Orion using the same Centaur III LOI stage as for the Lander. As the fully fueled Orion has a mass of 27 mT, there are potentially several tons of margin for this launch.
4) The Orion crew rendezvous with the Lander in LLO and transfers crew and possibly additional equipment and provisions enabled by the mass margin for the Orion launch.
5) The lander descends and lands out of LLO. The crew executes its surface mission, launches back to LLO in the ascent stage,

rendezvous with Orion, transfers crew, and deploys the ascent stage into a controlled lunar surface disposal.

6) The crew returns to Earth from LLO in Orion. The Orion ΔV capability of 1.25 km/s is more than sufficient for the trans-Earth insertion (TEI) maneuver.

LOI Stage

This stage is needed because the presently existing Orion service module ΔV capability of 1.25 km/s is sufficient for either insertion into or return to Earth from LLO, but not both. If developed for this purpose, it is likely to be advantageous to use the LOI stage also for insertion of the Lander into LLO. However, depending upon the efficiency of the Lander descent propulsion engine, it can be reasonable to consider making the Lander descent stage large enough to accommodate the additional, less-efficient, storable propellant necessary for insertion into LLO.

The present analysis does not incorporate this assumption. It is conservatively assumed here that the LOI stage will be used for both tasks and hence is sized for the more difficult requirement, Lander insertion into LLO. To this point, the fully fueled Centaur III with a single RL10C-1 engine (presently used as an upper stage for Atlas V) has the following parameters:

Specific Impulse (Isp) –	450 s
Dry Mass –	2.25 mT
Propellant Mass –	20.83 mT
Gross Mass –	23.08 mT
Diameter –	3.05 m
Length –	12.7 m

The required insertion ΔV from a three-day trans-lunar coast trajectory to LLO is approximately 1 km/s, depending in detail on a variety of factors including the choice of landing site. Assuming a required ΔV of 1 km/s for this analysis, the mass of propellant required for the Centaur III to insert the initial payload stack (Mi = 45 mT) into LLO is

$$Mp = Mi\,(1 - e\text{-}\Delta V/gIsp) = 9.2\ \text{mT}$$

and consists of about 8 mT of liquid oxygen and 1 mT of liquid hydrogen.

Propellant boiloff, primarily of the liquid hydrogen fuel, must be included in the cargo launch. For the production Centaur III, flown-vehicle data shows the loss rate to be 13-17% per day; with a few layers of insulation this can be reduced to 5% or less. (htps://www.google.com/url?sa=t&rct=j&q=&esrc=s&source=web&cd=&cad=rja&uact=8&ved=2ahUKEwj2hPPF5t-DaxWdF1kFHbJOBywQFnoECBMQAQ&url=htps%3A%2F%2Fwww.ulalaunch.com%2Fdocs%2Fdefault-source%2Fextended-duration%2Fcentaur-extensibility-for-long-duration-2006-7270.pdf&usg=AOvVaw1Vzv5kb-HhwZlszEs8dilI&opi=89978449). With this, the Centaur will lose less than 200 kg of propellant during a three-day trans-lunar coast. Including an allocation for a docking mechanism and other airborne support equipment for the Lander/Centaur cargo stack yields an allowable Lander mass of 32 mT, as shown:

SLS Block 2B TLI Payload	–	45 mT
Less		
Dry Mass, Centaur III	–	2.3 mT
LLO Insertion Propellant Mass	–	9.2 mT
Fuel Boiloff Allowance (5%/day)	–	0.2 mT
Airborne Support Equipment Allocation	–	1.3 mT
Subtotal for LOI Requirements	–	13 mT
Maximum Allowable Lander Mass	–	32 mT

The mass of the required LOI stage itself, slightly less than 12 mT, is about half the size of the Centaur III. For a lunar return mission, the stage could be flown as-is with a reduced propellant load, or a modified version with shorter tanks developed if desired. Also, the RL10C-1-1 engine variant for the Centaur V, the upper stage of Vulcan Centaur, offers an improved specific impulse of over 453 seconds. Given the time available before a lunar return mission will be executed, it may be feasible to incorporate this engine into a modified Centaur III LOI stage, thus gaining about 130 kg performance improvement.

Lunar Lander

To establish a baseline, the J-Series Apollo lunar landers (Apollo 15-17) had masses of less than 16.5 mT, including the 210 kg lunar rovers carried on each of these missions, and sustained two crewmembers for three days. Scaling of this experience would suggest that a four-person, 32 mT vehicle capable of supporting a 7-day mission is well within conservative design limits.

Improvements are possible, for example the incorporation of storable, low toxicity "green propellants" rather than the legacy, highly toxic, difficult to handle nitrogen tetroxide/hydrazine storable propellant combination. However, in the interest of offering a low schedule risk approach, the present analysis does not presume such advances.

Acquisition Strategy

The fundamental flaw in the Artemis acquisition approach is the assumption that the U.S. government can and should leverage so-called "commercial space" for nattional purposes, and that this paradigm is applicable to human spaceflight. It is debatable whether, in general, "commercial space" is other than a catchphrase intended to differentiate traditional prime contractors from newer firms aspiring to obtain government contracts without the excessive and shifling regulatory framework surrounding traditional government acquisition. However, it should be clear that no significant fiscal return on investment in human lunar missions can be expected in the foreseeable future without significant government subsidy.

It is thus NASA's responsibility to acknowledge that it is the only significant customer for human missions to the Moon and that it must therefore establish and direct a credible mission design to which contractors can bid, and to develop an equally credible cost estimate to implement that design, rather than agreeing to unrealistic firm fixed price (FFP) bids for complex development programs. Government FFP contracts that are underbid leave both sides stuck ina bad deal with only a few possible but unsatisfactory outcomes: the contractor demands additional money to finish the program and the government pays it, the program is ultimately canceled because the government doesn't want to pay, or performance is reduced in a compromise between the amount of money the contractor wants and that which the government is willing to pay. There is a long and depressing history of such efforts: (htps://www.defensenews.com/industry/2024/01/09/cautionary-tale-how-boeing-won-a-us-air-force-program-and-lost-7b/). We should not add human lunar return to the list.

If our nation is serious about returning to the Moon, this time to stay, then it properly requires an investment by the Congress on behalf of the public it serves. Congress and the public should expect that investment to be expertly managed by Executive Branch officials who are responsible and accountable for the quality of their decisions and the direction they provide to industry to implement those decisions. NASA's acquisition approach should reflect that fundamental principle.

Programmatic Considerations

The Artemis lunar landing missions as presently planned significantly compromise crew safety, carry high mission execution risk, are highly unlikely to remain on schedule, and are being executed via an inappropriate acquisition approach with grossly unrealistic fixed-price cost assumptions. These facts require hard decisions to be made if success is to be atained in the end. Congress must use its power of the purse to direct the Executive Branch to implement these decisions.

Or, we can just kick the can down the road, as we have been doing for more than five decades now.

Specifically, the existing contracts should be terminated for the convenience of the government and a new program initiated along the lines described above. Those who object will observe that termination for convenience will not allow significant funding to be recaptured from the existing fixed-price contracts, and this is correct. But to continue programs that we know will not achieve our goals distracts us from what must be done and damages NASA's and the nation's reputation, even if they are being executed for free. We need to focus our efforts on an approach that we know will work in a timely manner with the lowest mission risk and the greatest crew safety we can provide. To this point, while the analysis presented here offers a point design to illustrate concept feasibility, a sensitivity study should be conducted to establish the parametric feasibility space within which the two-launch mission design can be optimized.

Sustainability of our future space architecture does mater. Efforts to develop systems that expend fuel rather than hardware are important to that future. Because it is at the far end of the lunar ΔV gear train, a single-stage reusable crew lander is the most important of these developments. Thus, the development of cryogenic propellant transfer and zero-boiloff storage technologies should be pursued. But the development of these technologies will not be quick or easy, and timeliness is presently the more important feature for our nation's human lunar return program. Similarly, while NRHO and the Gateway as presently conceived are irrelevant to human lunar return, a transportation node or nodes like the Gateway will be valuable components of a sustainable future lunar architecture if placed in a more useful staging orbit than planned today.

But regardless of these finer points, the straightforward approach outlined here could put U.S.- led expeditions on the Moon beginning in 2029, given bold action by Congress and expeditious decision making and firm contractor direction by NASA. This is not the path being pursued at present and the existing Artemis contractual and programmatic structure will not support it. A new program, architected and managed by people who are clearly qualified for the job, should be initiated and executed with funding adequate to carry out this urgent and important national mission.

Michael Griffin Co-President

Michael D. Griffin is the Co-Founder and Co-President of LogiQ, Inc., a company providing high- end management, scientific, and technical consulting services. He was previously the Under Secretary of Defense for Research and Engineering, and in that role shared responsibility with the Deputy Under Secretary for research, development, and prototyping activities within the Department of Defense. In prior roles he was the Chairman and CEO of Schafer Corporation, the King-MacDonald Professor at University of Alabama in Huntsville, the Administrator of NASA, Space Department Head at the John Hopkins University Applied Physics Laboratory, President of In-Q-Tel, CEO of Magellan Systems, and EVP and General Manager of Orbital ATK's Space Systems Group. He is a member of the National Academy of

Engineering and the International Academy of Astronautics, an Honorary Fellow of the American Institute of Aeronautics and Astronautics, a Fellow of the American Astronautical Society, and a Fellow of the Institute of Electrical and Electronic Engineers. He has received the NASA Exceptional Achievement Medal, the AIAA Space Systems Medal and Goddard Astronautics Award, the National Space Club's Goddard Trophy, the Rotary National Award for Space Achievement, the Missile Defense Agency's Ronald Reagan Award, and has twice been awarded the Department of Defense Distinguished Public Service Medal. He holds seven earned degrees and has been recognized with honorary doctoral degrees from Florida Southern College and the University of Notre Dame. He is a Certified Flight Instructor with instrument and multiengine ratings, a Registered Professional Engineer in Maryland and California, and the lead author of two dozen technical papers and the textbook *Space Vehicle Design*.

Chapter 3

NASA Artemis Programs: Crewed Moon Landing Faces Multiple Challenges*

United States Government Accountability Office

Abbreviations

ACD	Artemis Campaign Development
ARGOS	Active Response Gravity Offload System
CCP	Commercial Crew Program
CDR	critical design review
EGS	Exploration Ground Systems
EHP	Extravehicular Activity and Human Surface Mobility Program
EVA	Extravehicular Activity
FAR	Federal Acquisition Regulation
HLS	Human Landing System
ISS	International Space Station
KDP	key decision point
NASA	National Aeronautics and Space Administration
Orion	Orion Multi-Purpose Crew Vehicle
PDR	preliminary design review
SLS	Space Launch System
TRL	technology readiness level

---

* This is an edited, reformatted and augmented version of the United States Government Accountability Office Report to Congressional Committees, Publication No. GAO-24-106256, dated November 2023.

In: Back to the Moon
Editor: Silas Blackwood
ISBN: 979-8-89530-861-5

Why GAO Did This Study

NASA is returning humans to the moon to maintain U.S. leadership in space exploration and prepare for future missions to Mars. NASA is implementing the Artemis missions to meet these goals. To accomplish the Artemis III mission as planned by December 2025, NASA needs to develop, acquire, and integrate several new systems. These include a system to transport crew to and from the lunar surface, and space suits for lunar surface operations.

NASA is using a relatively new approach to acquire the human landing system and space suits that is intended to increase innovation and improve affordability. To develop the lunar lander, NASA awarded a contract option to SpaceX in 2021.

To develop Artemis space suits, it awarded a contract to Axiom Space in 2022.

A House report includes a provision for GAO to review NASA's lunar programs. This is GAO's fourth report examining the Artemis enterprise.

This report describes the extent to which NASA has made progress in developing key systems needed to land humans on the moon in 2025, and has processes in place to ensure that those systems will meet NASA's needs and be safe.

GAO assessed NASA data, documentation, and policy; analyzed contract documentation, contractor risk charts, and technology maturation plans; and interviewed NASA officials and industry representatives.

What GAO Found

The National Aeronautics and Space Administration (NASA) is preparing to land humans on the moon for the first time since 1972 in a mission known as Artemis III. Since GAO's September 2022 report (GAO-22-105323), NASA and its contractors have made progress, including completing several important milestones, but they still face multiple challenges with development of the human landing system and the space suits. As a result, GAO found that the Artemis III crewed lunar landing is unlikely to occur in 2025. In July 2023, NASA stated that it is reviewing the Human Landing System schedule.

The current challenges that GAO identified include:

- *An ambitious schedule:* The Human Landing System program is aiming to complete its development—from project start to launch—in 79 months, which is 13 months shorter than the average for NASA major projects. The complexity of human spaceflight suggests that it is unrealistic to expect the program to complete development more than a year faster than the average for NASA major projects, the majority of which are not human spaceflight projects. GAO found that if development took as long as the average for NASA major projects, the Artemis III mission would likely occur in early 2027.
- *Delays to key events:* As of September 2023, the Human Landing System program had delayed eight of 13 key events by at least 6 months. Two of these events have been delayed to 2025—the year the lander is planned to launch. The delays were caused in part by the Orbital Flight Test, which was intended to demonstrate certain features of the launch vehicle and lander configuration in flight. The test was delayed by 7 months to April 2023. It was then terminated early when the vehicle deviated from its expected trajectory and began to tumble. Subsequent tests rely on successful completion of a second Orbital Flight Test.

Notional Depiction of the Human Landing System

Source: SpaceX. | GAO-24-106256.

- *A large volume of remaining work:* SpaceX must complete a significant amount of complex technical work to support the Artemis

III lunar landing mission, including developing the ability to store and transfer propellant while in orbit. A critical aspect of SpaceX's plan for landing astronauts on the moon for Artemis III is launching multiple tankers that will transfer propellant to a depot in space before transferring that propellant to the human landing system. NASA documentation states that SpaceX has made limited progress maturing the technologies needed to support this aspect of its plan.

- *Design challenges:* Axiom is leveraging many aspects of NASA's prior work to develop modernized space suits, but significant work remains to resolve design challenges. For example, NASA's original design did not provide the minimum amount of emergency life support needed for the Artemis III mission. As a result, Axiom representatives said they may redesign certain aspects of the space suit, which could delay its delivery for the mission.

Illustration of Axiom's Space Suit and Major System Components

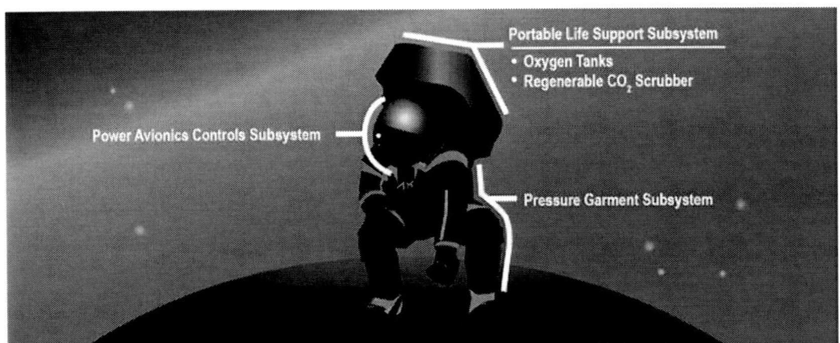

Source: GAO analysis of Axiom information and image. | GAO-24-106256.

NASA plans to take multiple steps to determine whether SpaceX's and Axiom's systems meet its mission needs and are safe for crew. For example, NASA developed a supplemental process—one not required by its policies—to determine whether the contractors' systems meet requirements before the mission. Also, NASA's contracting approach to acquire the human landing system and space suits as services included insight clauses in the SpaceX and Axiom contracts. Program officials stated these clauses ensure that NASA has visibility into broad aspects of the contractors' development work, including anything that could affect the Artemis III mission or crew safety. Officials stated that this visibility extends to certain aspects of work SpaceX and Axiom are doing for their commercial endeavors. For example, this included SpaceX's activities leading up to the Orbital Flight Test, which flew a commercial variant of the human landing system.

November 30, 2023

The Honorable Jeanne Shaheen
Chair

The Honorable Jerry Moran
Ranking Member
Subcommittee on Commerce, Justice, Science, and Related Agencies
Committee on Appropriations
United States Senate

The Honorable Hal Rogers
Chairman

The Honorable Matt Cartwright
Ranking Member
Subcommittee on Commerce, Justice, Science, and Related Agencies
Committee on Appropriations
House of Representatives

The National Aeronautics and Space Administration (NASA) plans to return U.S. astronauts to the surface of the moon by the end of 2025. This mission, known as Artemis III, will be the first time that crew will land on the moon since the 1972 Apollo 17 mission, and the first time ever that crew will land at the lunar south pole. The Artemis III mission is the third in a series of increasingly complex missions to maintain U.S. leadership in space exploration, build a sustainable lunar presence over the next decade, and ultimately travel to Mars.

In March 2019, the White House directed NASA to accelerate its plans for a lunar landing from its original goal of 2028 to 2024, in part to create a sense of urgency in returning American astronauts to the moon. In November 2021, NASA announced that it was no longer working to its goal of an Artemis III lunar landing in 2024 and that the new date would be no earlier than 2025. NASA officials attributed this change to a 7- month delay in working on the human landing system, subsequent to a bid protest and federal court complaint regarding the award of the lander's contract. In announcing the delay, senior NASA officials acknowledged that the prior 2024 goal was unrealistic.

To accomplish the Artemis III mission, NASA is partnering with industry to develop two new systems: the human landing system (HLS)—to transport crew to and from the lunar surface, and modernized space suits for lunar surface operations.[1] In the fiscal year 2024 President's budget request, NASA requested $12.4 billion over the next 5 fiscal years for the human landing system and modernized space suits. In addition to developing these systems, NASA will need to ensure that the human landing system and space suits meet NASA's needs to operate in a deep space environment, conduct scientific exploration, and ensure crew safety.

The House Report 117-395 accompanying the Commerce, Justice, Science, and Related Agencies Appropriations bill, 2023 contains a provision for GAO to continue conducting in-depth reviews of NASA's lunar-focused programs. This is the fourth in a series of GAO reports addressing NASA's Artemis enterprise.[2] The focus of this report is on the human landing system and space suits being developed for the Artemis III mission. This report describes the extent to which NASA (1) has made progress in developing key systems needed to land humans on the moon in 2025, and (2) has processes in place to ensure that its contractors are developing systems that meet NASA mission needs and are safe for crew.

To conduct this work, we reviewed and assessed NASA data, documentation, and policy. For example, we reviewed program plans and quarterly status reviews from the HLS and Extravehicular Activity and Human Surface Mobility (EHP) programs, which are overseeing the development of the human landing system and modernized space suits, respectively. We reviewed these documents to identify program milestones, critical technology demonstrations, and NASA's progress. We also analyzed contract documentation, contractor risk charts, and technology maturation plans to determine the current risks facing the programs. Additionally, we assessed

---

[1] NASA uses the term Exploration Extravehicular Activity system to encompass the garments, interfaces with the human landing system, and tools for the scientific lunar mission. For the purposes of this report, we will refer to the Exploration Extravehicular Activity system as space suits and associated tools.

[2] GAO, *NASA Lunar Programs: Improved Mission Guidance Needed as Artemis Complexity Grows*, GAO-22-105323 (Washington D.C.: Sept. 8, 2022); *NASA Lunar Programs: Significant Work Remains, Underscoring Challenges to Achieving Moon Landing in 2024*, GAO-21-330 (Washington, D.C.: May 26, 2021); GAO-21-105; and *NASA Lunar Programs: Opportunities Exist to Strengthen Analyses and Plans for Moon Landing*, GAO-20-68 (Washington, D.C.: Dec. 19, 2019). For more information, see our related work at the end of this report.

contract and requirements documentation, NASA agendas and presentation slides on lessons learned, and mission integration documentation. We also interviewed a wide range of NASA and industry officials. See appendix I for more information on our objectives, scope, and methodology.

We conducted this performance audit from September 2022 to November 2023 in accordance with generally accepted government auditing standards. Those standards require that we plan and perform the audit to obtain sufficient, appropriate evidence to provide a reasonable basis for our findings and conclusions based on our audit objectives. We believe that the evidence obtained provides a reasonable basis for our findings and conclusions based on our audit objectives.

Background

NASA's Artemis Missions

The goal of NASA's Artemis enterprise is to return U.S. astronauts to the surface of the moon, establish a sustained lunar presence, and ultimately achieve human exploration of Mars. To do so, NASA programs are developing multiple highly complex and interdependent systems that will need to be integrated to support individual Artemis missions.[3]

- The Artemis I and II missions are the first uncrewed and crewed demonstration missions, respectively, of the Space Launch System (SLS) launch vehicle, Orion Multi-Purpose Crew Vehicle (Orion), and associated ground systems, known as Exploration Ground Systems (EGS).[4] Artemis I successfully launched on November 16, 2022, with the Orion capsule safely returning to Earth on December 11, 2022.
- The Artemis III mission incorporates new programs that are developing the human landing system and space suits. The goal of the

[3] NASA distinguishes between programs and projects in its policies and guidance. A NASA program has a dedicated funding profile and defined management structure and may include several projects. Projects are specific investments under a program that have defined requirements, life-cycle costs, schedules, and their own management structures.

[4] SLS is the vehicle NASA will use to launch the Orion Crew Capsule beyond low-Earth orbit. Orion is the crew capsule to transport humans from the Earth to the HLS. EGS is the infrastructure at Kennedy Space Center to support Artemis mission launches.

Artemis III mission is to return U.S. astronauts to the surface of the moon and conduct scientific exploration activities during a 6.5-day stay at the lunar south pole. The characteristics of the lunar south pole affect when and where crew can land. For example, the mission must occur in a location that stays continuously lit, and the crew must have direct-with-Earth communication.

- For Artemis IV and beyond, NASA is also developing a lunar orbiting outpost known as the Gateway to act as a habitat and safe work environment for astronauts and as a communications relay between the lunar surface and Earth.

See figure 1 for the programs needed to accomplish the Artemis missions.

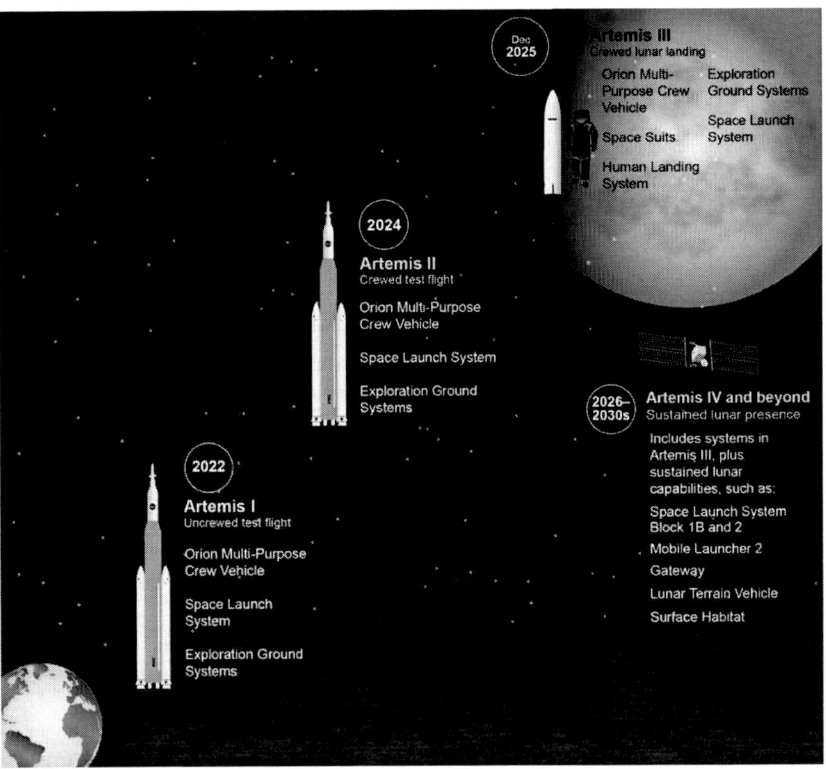

Source: GAO pesentation of NASA documentation. | GAO-24-106256.

Figure 1. Key NASA Programs Supporting Artemis Missions.

Key NASA Program Offices Supporting the Artemis III Mission

Several key program offices have a role in supporting the Artemis III mission.

- The HLS program is responsible for managing the human landing system development and certifying that the integrated lander systems are ready for flight. The program management is located at Marshall Space Flight Center, with key participation from Johnson Space Center and Kennedy Space Center.
- EHP is responsible for working with industry to advance technologies associated with human mobility and lunar surface infrastructure to support the Artemis missions. The Extravehicular Activity (EVA) Development Project within EHP manages the space suit development for the Artemis missions and is located at the Johnson Space Center.[5]
- In March 2023, NASA established the Moon to Mars program office to oversee the programs contributing to the Artemis missions, including SLS, Orion, EGS, HLS, space suits, and the Gateway. This new program is responsible for end-to-end risk management and risk acceptance for the entire exploration system. The new program office resides within the Exploration Systems Development Mission Directorate. Prior to establishing the Moon to Mars program and during the course of our review, the Artemis Campaign Development (ACD) Division was responsible for overseeing the Artemis III mission integration. The Moon to Mars program is leveraging previous work performed by ACD on integrating across the programs to support the Artemis missions.

Contracting Approach for Human Landing System and Space Suit Development

To support the Artemis III mission, NASA awarded firm-fixed-price indefinite delivery, indefinite quantity contracts to two companies, SpaceX and Axiom

[5] The EVA Development project also manages space suit development for the International Space Station, which is outside the scope of this report. For the purposes of this report, references to EHP and its officials encompass the EVA Development project.

Space.[6] SpaceX is to develop and demonstrate the human landing system, and Axiom Space is to develop the space suits. In July 2021, NASA exercised a $2.9 billion option on its contract with SpaceX to provide crew access to the lunar surface and demonstrate initial capabilities required for deep space missions.[7] Subsequently, in May 2022, NASA awarded firm-fixed-price indefinite delivery, indefinite quantity contracts to Axiom Space and Collins Aerospace. These companies are to provide safe and reliable commercial extra vehicular activity services in microgravity and partial gravity environments on the International Space Station and the lunar surface for Artemis missions.[8]8 In September 2022, NASA issued a $229 million order under Axiom's contract for the development and demonstration of a suit for lunar surface activities. Axiom is required to provide space suits that will allow crew to successfully perform exploration and science missions on the lunar surface during the Artemis III mission.

NASA is acquiring the human landing system and space suits as services, which represents a relatively new contracting approach for NASA.[9] NASA's intent is to transition from the traditional government- owned hardware model to a service model. Using this approach, the programs set high-level requirements and rely on the contractor's innovation to develop and deliver the system. Contractors only receive payment after NASA determines that the contractor has successfully achieved a milestone as defined in the contract.

---

[6] Under a firm-fixed-price-type contract, the government pays a fixed price regardless of actual cost and places upon the contractor maximum risk and full responsibility for all costs and resulting profit or loss. A firm-fixed-price contract provides for a price that is not subject to any adjustment on the basis of the contractor's cost experience in performing the contract. It provides maximum incentive for the contractor to control costs and perform effectively and imposes a minimum administration burden upon the contracting parties. FAR 16.202-1.

[7] NASA first awarded the HLS contract to three providers in May 2020. In April 2021, NASA announced the selection of SpaceX for the award of the contract to develop the Artemis III human landing system. After the award, Blue Origin and Dynetics filed bid protests with GAO, which GAO denied in July 2021. GAO, *Blue Origin Federation, LLC; Dynetics, Inc.-A Leidos Company,* B-419783; B-419783.2; B-419783.3; B-419783.4, July 30, 2021, 2021 ¶ CPD 265 (Washington, D.C.: July 30, 2021). Subsequently, in August 2021, Blue Origin filed a complaint with the U.S. Federal Court of Claims. The court dismissed this complaint in November 2021. Blue Origin Fed. LLC v. United States, Fed. Cl., No. 21-1695C (Nov. 4, 2021).

[8] An indefinite delivery, indefinite quantity contract provides for an indefinite quantity, within stated limits, of supplies or services during a fixed period. The government places orders for individual requirements. FAR 16.504(a).

[9] The acquisition strategy for the HLS program involves procurement of initial landing capabilities from each provider as Research and Development per FAR Part 35. Recurring services per FAR Part 37 will be procured in later acquisitions.

NASA is moving toward using this service model contracting approach for some acquisitions on its Artemis missions because the agency believes that it increases competition, innovation, flexibility, speed, and affordability.

Previously, NASA used a service model contract approach for its Commercial Crew Program (CCP). We reported in 2019 that NASA awarded firm-fixed-price contracts to SpaceX and Boeing in 2014 to acquire human space flight transportation services to and from the ISS for CCP.[10] Under the contracts, we reported NASA also evaluated whether the contractors met its requirements and certified contractor final systems for use. In 2020, NASA determined that one of the CCP contractors, SpaceX, met the agency's standards for human space flight and certified it to conduct crewed missions to and from the ISS. As of May 2023, the other contractor, Boeing, was still working toward certification.

Prior to awarding the contracts for the human landing system and space suits, NASA took steps to reduce risk and increase industry participation in the programs. For example,

- To inform the human landing system development, NASA conducted risk reduction studies on topics including cryogenic fluid management and landing systems. Through these studies, NASA worked with several companies to get feedback on proposed human landing system requirements, develop element designs, and identify key technologies.
- NASA made its space suit design available to potential contractors and published a technical library with details of the design on the NASA website. NASA documentation states that the agency made this information available to reduce cost, schedule, and technical risks and contractor barriers to entry for providing a space suit for the Artemis III mission.
- The HLS program and EHP sought lessons learned from prior commercial human space flight programs, including CCP, to inform their acquisition strategies for the human landing system and space suits. Between 2019 and 2022, multiple NASA centers and mission directorates hosted knowledge-sharing events to facilitate the sharing

[10] GAO, NASA Commercial Crew Program: Schedule Uncertainty Persists for Start of Operational Missions to the International Space Station, GAO-19-504 (Washington, D.C.: June 20, 2019).

of lessons learned from programs that used commercial service-type contracts. HLS and EHP officials attended these events.

Key Elements of the Artemis III Human Landing System

The human landing system will provide crew access to the lunar surface and demonstrate initial capabilities required for deep space missions.

SpaceX is currently developing a commercial Starship vehicle to transport humans and cargo to low-Earth orbit, the moon, and Mars.

The HLS Starship system consists of the SpaceX Super Heavy booster (launch vehicle) and HLS Starship (the vehicle that provides crew access to the lunar surface). The HLS Starship is based on a common Starship architecture and shares many of the same critical systems, including propulsion, structures, and avionics, with the commercial Starship. The Raptor engines—which require liquid methane and liquid oxygen (collectively referred to as propellant)—power both the Super Heavy booster and the HLS Starship. In addition to the HLS Starship, SpaceX is developing a propellant tanker and on-orbit propellant depot—also variants of the commercial Starship vehicle—for its lunar landing mission concept.

SpaceX's plan for landing NASA astronauts on the moon for Artemis III includes multiple steps conducted sequentially. The order of these steps is as follows:

1) The propellant depot will be launched to low-Earth orbit, followed by multiple tankers that will rendezvous with, dock to, and transfer propellant to the depot;
2) Once sufficient propellant is on-orbit, an uncrewed HLS Starship will launch into low-Earth orbit, then rendezvous with and dock to the depot. The depot will transfer its propellant to the HLS Starship. The HLS Starship will then perform a rapid transfer into near-rectilinear halo orbit, where it will loiter for up to 90 days to confirm vehicle health and await the launch and arrival of Orion (the 90-day time frame is to accommodate any potential Orion or SLS launch delays);[11]
3) Orion will then launch with crew on board and dock with the HLS Starship;

[11] Near-rectilinear halo orbit is a 1-week lunar orbit balanced between the Earth's and moon's gravity. This orbit enables global lunar access and promotes access to the lunar poles.

4) Two astronauts will transfer from Orion into the HLS Starship, which will descend to the lunar surface for a 6.5-day stay; and,
5) Once the lunar surface activities, including moonwalks, are complete, the HLS Starship will ascend back to near-rectilinear halo orbit, where the crew will transfer back to Orion for their return to Earth.

Figure 2 depicts SpaceX's mission concept for the Artemis III lunar landing.

Key Elements of the Artemis III Space Suits

Modernized space suits and associated hardware will provide portable life support, as well as tools for crew to use for lunar science and maintenance tasks. In December 2021, after spending 14 years and $420 million on next generation space suit development, NASA determined it would acquire modernized space suits from industry rather than develop the space suit system in-house.[12] NASA worked to mature several aspects of its design and, as noted above, made that design available publicly to help enable industry development of a space suit system.

NASA's design is known as the government reference design.

To deliver and demonstrate lunar surface space suits and associated systems, Axiom is leveraging many aspects of NASA's government reference design. The space suit consists of a combination of a pressure garment and life support components that together will provide capacity for at least 8 hours of lunar surface activity. The in-space system will consist of a variety of tools, flight support equipment, and other hardware that enable lunar surface exploration activities.

Life Cycle of a NASA Space Flight Project

NASA is using its life-cycle review process to oversee the contractors' development of the human landing system and the space suits. This process consists of two phases— (1) formulation, which takes a project from concept development to preliminary design, and (2) implementation, which includes activities like building, launching, and operating the system. Major projects must get approval from senior NASA officials at key decision points (KDP) before they can enter each new phase. Figure 3 depicts NASA's life cycle for space flight projects.

---

[12] NASA Office of Inspector General, *NASA's Development of Next-Generation Spacesuits*, IG-21-025 (Washington, D.C.: Aug. 10, 2021).

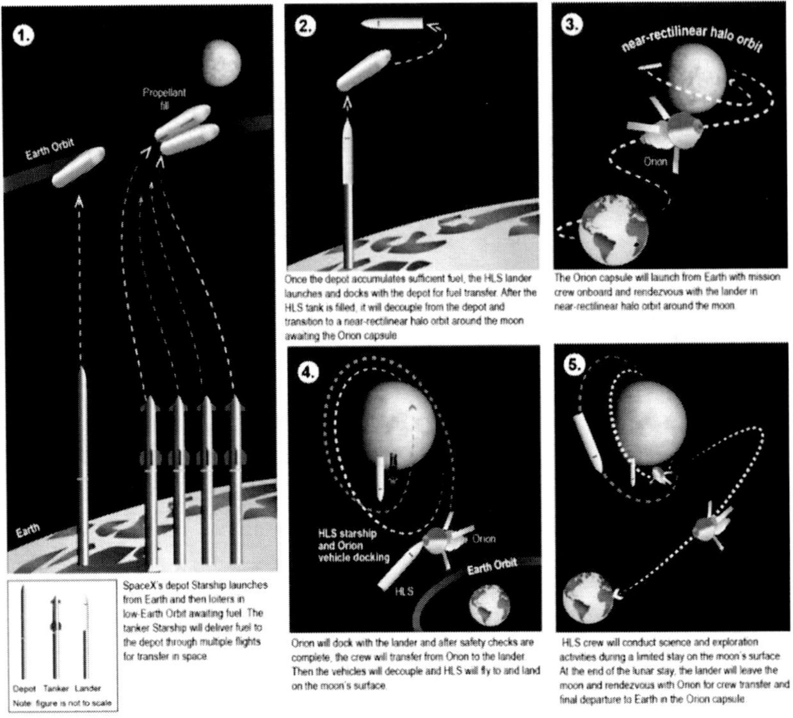

Source: GAO presentation of NASA and SpaceX information. | GAO-24-106256.

Figure 2. SpaceX Mission Concept for Human Landing System (HLS).

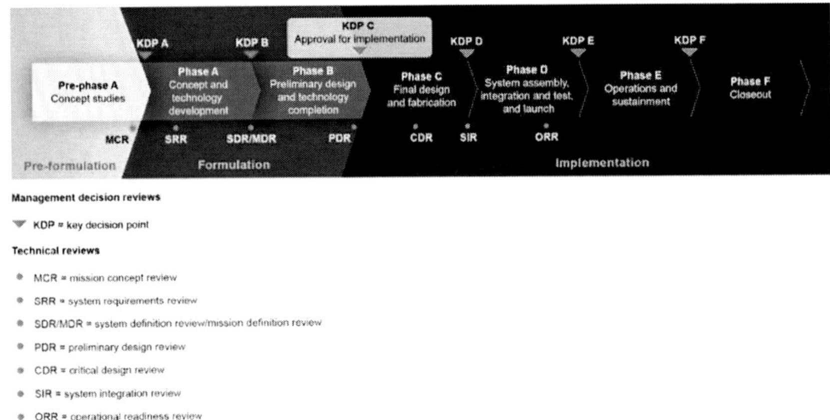

Source: GAO presentation of NASA images and information. | GAO-24-106256.

Figure 3. NASA's Life Cycle for Space Flight Projects.

In phase A, a project team develops a range of cost and schedule estimates for uses such as budget planning. During phase B, the project team develops programmatic measures and technical leading indicators that track various project metrics such as requirements changes, staffing demands, and mass and power utilization. Near the end of formulation, leading up to the preliminary design review (PDR), the project team completes technology development and its preliminary design.

Formulation culminates in a review at KDP C, where senior leaders approve the cost and schedule agency baseline commitments. After a project holds KDP C, it begins implementation, consisting of phase C where the project team holds a critical design review (CDR) to determine whether the design performs as expected and is stable enough to support proceeding with the final design and fabrication. As of July 2023, the HLS program is approaching KDP C while the EVA Development project—i.e., the space suit project—is approaching PDR.

2025 Crewed Lunar Landing Is Jeopardized by Delays and Scale of Remaining Work

NASA and SpaceX completed several important milestones for the human landing system since our September 2022 report, but a variety of factors make a lunar landing in 2025 unlikely. NASA officials are currently reviewing the HLS schedule. Axiom is making progress on the space suits, but it also has significant work to complete before the planned 2025 launch. At the same time, NASA and the HLS and Extravehicular Activity and Human Surface Mobility (EHP) programs are addressing a wide range of cross-program risks to landing humans on the moon in 2025.

Limited Lander Progress Makes 2025 Launch Unlikely Given Delays and Remaining Work

SpaceX has made progress in designing and testing components of its HLS Starship which, as discussed above, is the human landing system that will provide crew access to the lunar surface. However, the contractor is facing multiple issues that limit this progress and jeopardize its ability to support an Artemis III mission in 2025. These issues include an ambitious schedule, delays to key events, an incomplete orbital flight test, and a large volume of remaining technical work.

Ambitious Schedule

We found that if the HLS development takes as many months as NASA major projects do, on average, the Artemis III mission would likely occur in early 2027. Our analysis of past NASA projects that have launched since 2010 found that the average development time from project start to launch was 92 months. NASA has already delayed the Artemis III mission to December 2025, extending the HLS development time to 79 months.

However, this is still 13 months faster than the average development time for NASA major projects. The complexity of human spaceflight suggests that it is unrealistic to expect the HLS program to complete development more than a year faster than the average for NASA major projects, the majority of which are not human spaceflight projects.

While SpaceX and NASA are aiming to complete development more than a year faster than the average for NASA major projects, they are achieving key events at a slower pace. For example, we found that SpaceX used more than 50 percent of its total schedule to reach PDR in November 2022. On average, NASA major projects used about 35 percent of the total schedule to reach this milestone.

Furthermore, the HLS program is taking longer to reach KDP C—the next key review after PDR—than average for the NASA major projects we assessed. As a result, the HLS program is proceeding with development without formal approval of a cost and schedule baseline. Specifically, the HLS program plans to use nearly 14 percent more of its total schedule to proceed from PDR to KDP C while on average NASA major projects used just 4.2 percent more of their schedule to achieve KDP C.[13]

Figure 4 illustrates the percent of total schedule used in achieving PDR and the percent that will be used in achieving KDP C by the HLS program, compared to the average for a NASA major project in our dataset.

Delays to Key Events

SpaceX has also delayed several future program events that further compress the schedule. Since July 2022, the HLS program office and SpaceX delayed multiple key events from 2023 to 2024, meaning that many critical

[13] For more information on how we completed this analysis, see appendix I.

demonstrations and reviews will need to occur in the next 2 years to support an Artemis III mission as planned in 2025.

SpaceX and NASA continue to make progress on the human landing system, including completing some work early. SpaceX representatives reported completing 20 interim HLS milestones since June 2022 to mature the human landing system design and reduce development risk. NASA officials stated SpaceX submitted deliverables early for approximately 74 percent of the Artemis III contract payment milestones that have been completed.

Overall, the HLS program and SpaceX delayed eight out of 13 key events by between 6 and 13 months. Of those delayed events, at least two will occur in 2025—the year the Artemis III mission is scheduled to take place. Partially as a result of these delays, SpaceX plans to complete eight key events between November 2023 and the planned date of the Artemis III mission.

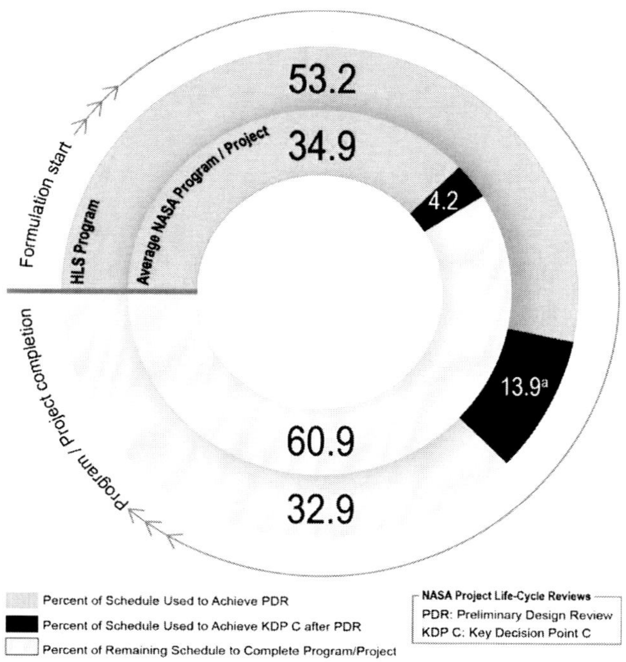

Source: GAO analysis of NASA data | GAO-24-106256.

[a] This calculation assumes the KDP C review will occur in October 2023, as currently planned, but that date was still in the future as of the writing of this report.

Figure 4. Human Landing System (HLS) Program Used a Greater Schedule Percentage to Achieve Planned Key Reviews than the Average for NASA Major Projects Launched since 2010.

Due to delays to several key events, NASA will have a relatively short amount of time to ensure that the HLS complies with human spaceflight safety requirements before the mission start. For example, NASA delayed the HLS Design Certification Review, which is now closer to the Artemis III mission than originally planned. At this review, NASA will ensure that the design complies with requirements and human spaceflight certification. According to NASA documentation, this milestone should be completed 9 months prior to launch. As of September 2023, NASA and SpaceX also planned to complete Flight Readiness Reviews for the depot, tanker, and lander versions of the Starship within the same 9- month period before Artemis III. NASA and SpaceX will have less time to address any issues identified during these reviews before the mission.

SpaceX representatives told us the three flight readiness reviews will always occur within the same length of time to support the Artemis III mission due to the nature and cadence of reviews. Any additional delays to these key events will compound the schedule pressure on NASA and SpaceX.

Incomplete Orbital Flight Test

In April 2023, after a 7-month delay, SpaceX achieved liftoff of the combined commercial Starship variant and Super Heavy booster during the Orbital Flight Test. But, according to SpaceX representatives, the flight test was not fully completed due to a fire inside the booster, which ultimately led to a loss of control of the vehicle. Following the launch, the Federal Aviation Administration—which issues commercial launch and re- entry licenses—classified the commercial Starship launch as a mishap and required SpaceX to conduct a mishap investigation. The Federal Aviation Administration reviewed the August 2023 mishap report submitted by SpaceX and, as a result, cited 63 corrective actions for SpaceX to implement before a second test.

SpaceX had planned this demonstration as the first test flight of the booster stage, as well as the first test with the Starship riding on the booster and the whole system experiencing stage separation.[14] However, SpaceX representatives said their Autonomous Flight Safety System initiated the vehicle self-destruct sequence and the vehicle began to break up about 4 minutes into the flight after the vehicle deviated from the expected trajectory, lost altitude, and began to tumble. HLS officials said that while the flight test was terminated early, it still provided data for several Starship technologies,

[14] Stage separation occurs when the Super Heavy booster (first stage) and HLS Starship (second stage) disconnect after launch.

including propellant loading, launch operations, avionics, and propulsion behavior.

The incomplete Orbital Flight Test led NASA to delay many key test events that are dependent on completion of that test. For example, NASA officials said that the in-space propellant transfer test will be delayed because it requires SpaceX to demonstrate that the Starship vehicle can reach orbit. Likewise, HLS officials told us that the Starship tests are sequentially linked, so future test flights, including the uncrewed flight test, depend on SpaceX successfully completing both the Orbital Flight and in-space propellant transfer tests.

A successful Orbital Flight Test is also needed to execute the technology maturation plan for the human landing system. For example, according to SpaceX representatives, the April 2023 Orbital Flight Test was key to demonstrating and understanding many aspects of the launch operations and ascent performance. SpaceX representatives said that they collected early in-flight data on the Raptor engines, vehicle tanks and primary structures, and pad and ground systems from the Orbital Flight Test.

However, HLS officials stated that SpaceX is still required to perform a successful booster separation, ignite the Starship's engines, and achieve a suborbital altitude. HLS officials said that reaching orbit is essential for Starship development and that they expect that SpaceX's pace of design changes is likely to increase after a successful test flight, allowing them to make progress on finalizing the design and building hardware. SpaceX documentation states it plans to fly a second test in the fourth quarter of 2023.[15]

Lastly, NASA program officials said it is unknown whether the HLS Starship can be ready by the December 2025 launch date. The HLS program schedule required adjustments after the incomplete Orbital Flight Test and subsequent mishap investigation. In July 2023, NASA provided documents stating that the HLS schedule, including the program's ability to support a December 2025 launch date, are under review.

Remaining Technical Work

The HLS program will need to complete a significant amount of complex technical work on the engines and propellant transfer technology between 2023 and the end of 2025 to achieve the planned lunar landing goal. In a May 2023 NASA document, NASA officials overseeing the Artemis III mission

---

[15] SpaceX conducted the second Orbital Flight Test on November 18, 2023, and, this test was outside the scope of our review.

integration stated that the HLS design maturity, with almost 3 years left before the planned launch, was insufficient. For example, they cited on-orbit propellant transfer as a potential issue because significant technical problems still need to be resolved.

The remaining technical work includes:

- Raptor engine development. SpaceX plans to use the Raptor engine in both the lander and booster stages of the human landing system and considers the technology to be relatively mature because it incorporates many years of prior development. However, the HLS Program Office identified engine development as a top risk to the program. SpaceX representatives said that their design for the Raptor engine follows an iterative approach, and as of September 2023, SpaceX had assembled and tested hundreds of engines. In a February 2023 interview, HLS officials said that if the Raptor engine operates below performance levels needed to meet mission requirements, thereby delaying engine certification, then it is possible that the new main engine for the human landing system will not be ready to support the planned mission in December 2025.
- On-orbit propellant transfer technology. SpaceX has remaining technical work to develop its on-orbit propellant storage and transfer technology. HLS program documentation states that propellant storage and transfer technologies have not previously flown in an integrated propulsion-like system. The documentation noted that, to date, SpaceX has made limited progress in maturing those technologies.

There are multiple key systems related to the propellant transfer capability that SpaceX plans to develop for the human landing system. Those systems include docking sensors and mechanisms (to identify, locate, and then physically align the HLS Starship and the tanker Starship for fluid transfer); propellant measurement (to gauge the amount of propellant in the tanks and how much was transferred); and storage capability to mitigate fuel loss in space.

SpaceX plans to conduct the in-space Propellant Storage and Transfer test to further mature the technology, but the timing of this test is dependent on successful completion of preceding flights.

SpaceX representatives said that the fundamental propellant transfer technology is not new or unique but requires engineering time and development effort to fully design and test for eventual use in the Artemis III mission. If the docking hardware does not perform as expected during spaceflight testing, significant vehicle modifications may be required, which could delay the mission. HLS officials said that SpaceX must demonstrate these technologies prior to completing the critical design review to promote confidence in the mission concept of operations.

Axiom Is Making Progress on Space Suits but Significant Work Remains

Axiom made progress in developing the space suits by completing several milestones, but it is still in the early phases of development. Since NASA's EHP gave Axiom authorization to proceed in September 2022, Axiom completed two NASA life-cycle milestones—Mission Concept Review in December 2022 and Certification Baseline Review in March 2023.[16]

In December 2022, Axiom built the first engineering development suit, derived from NASA's government reference design, with Axiom- constructed components. Axiom representatives said they use this development suit to test and redesign different components. In December, Axiom tested the suit at operational pressure in the lab using employees as in-house subjects to go through mobility exercise tests.

Axiom also used the development suit to assess the sizing approach on test subjects. EHP officials said that Axiom is on schedule to complete PDR in November 2023. Figure 5 shows Axiom's completed and remaining life-cycle milestones to deliver space suits to NASA for the planned December 2025 launch.

Axiom is also making progress on suit development by leveraging NASA's prior work. Axiom representatives said they brought in relevant experts and personnel who worked on the government reference design. These representatives said their approach was to adopt the government reference design and refine it to reduce costs to NASA.

At a February 2023 program review meeting with NASA, Axiom discussed its planned modifications to the government reference design space suit and its approach to mature several subsystem components in preparation

[16] The Certification Baseline Review is the milestone when the contractor establishes the design baseline, certification plan, life-cycle costs, and schedules for system certification. This milestone occurs early in the development phase and prior to PDR.

for PDR in November 2023. For example, Axiom is repackaging the life support system to increase the size of the suits' oxygen tanks and building new components to improve NASA's government reference design. See figure 6 for an illustration of critical space suit subsystems and components that Axiom must mature by the planned December 2025 mission.

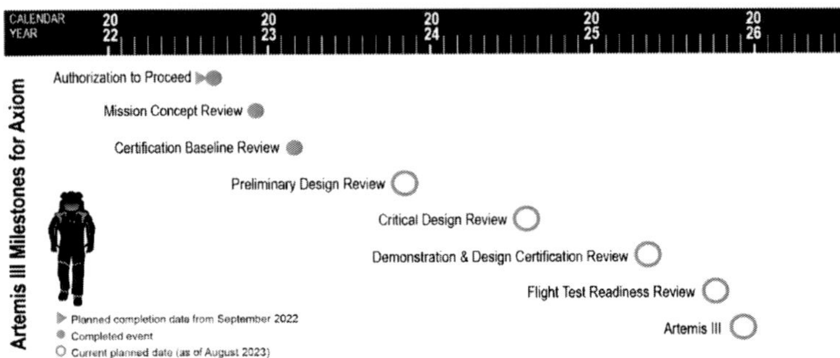

Source: GAO analysis of NASA and Axiom information/ | GAO-24-106256.

Figure 5. Completed and Remaining Milestones to Develop Space Suits for the Artemis III Mission.

Source: GAO analysis of NASA and Axiom information/ | GAO-24-106256.

Figure 6. Illustration of Axiom's Space Suit and Major System Components.

In addition, Axiom is modifying certain components of the government reference design to meet challenging requirements and address parts obsolescence issues.[17]

- *Emergency life support requirements.* NASA is requiring Axiom to develop a suit that can provide 60 minutes of emergency life support, more than any suit in history. Axiom representatives told us they may redesign applicable portions of the suit because the NASA government reference design did not satisfy the requirement to make the suit capable of storing that amount of oxygen. Axiom staff plan to decrease the size and rearrange components in the life support system package design to accommodate larger tanks that can hold more oxygen. However, if that is not possible, they plan to modify the current life support system, which Axiom representatives told us will take additional time.
- *Parts obsolescence and improvements.* Axiom representatives said they plan to incorporate, design, and certify new technologies— i.e., batteries, pumps, and electronic components—because designing new components helps them address supply chain and obsolescence issues. For example, Axiom is designing their space suit to be modular so that new technologies can be incorporated to address mission-specific requirements and allow incremental system upgrades. Axiom's remaining technical work on the space suits include modifications to all three sub-systems (Power, Avionics, and Controls; Pressure Garment System; and Life Support System).

  Axiom is also making several improvements to the government reference design, which may require additional testing to mature. For example, Axiom considers one of the Axiom-designed life support components, the heat exchanger, as less mature than other suit components. However, Axiom representatives said their heat exchanger is outperforming the government reference design based on their testing. Additionally, NASA officials said Axiom is developing and building its own evaporative fibers for the space suit water membrane evaporator—a key system for cooling the suit's

[17] EHP officials said that the government reference design had limitations and challenges because it was not finished when NASA shifted strategies for providing space suits for the Artemis III mission. NASA officials said that the reference design suit had limitations because it was only matured to the CDR stage. At CDR, the design is mature enough to support full-scale fabrication, assembly, integration, and testing, but is still incomplete.

temperature. NASA officials said Axiom reported that modifications to the evaporative fibers are performing similarly, if not better, than the government reference design. NASA requested some of the new fibers Axiom proposed for the design to perform its own assessment but has not completed the assessment. As such, these component modifications could require further testing to mature.

In addition to resolving design issues for the space suits, Axiom has significant other remaining work to complete within the next 2 years:

- *Mature critical technologies.* Axiom plans to mature several critical technologies to meet mission requirements. For example, in January 2023, NASA assessed two critical space suit systems, the Life Support System and the Pressure Garment System, at a technology readiness level (TRL) 4.[18] Axiom's technology assessment rated over half of its critical technologies below TRL 6 and the lowest among those items at a TRL 3. One of those components, the Regenerable CO2 Scrubber, is a life support system subcomponent that removes carbon dioxide from the suit environment. Axiom rated it at a low TRL since Axiom is not using the government reference design for this component.

  NASA officials told us they expect critical technologies to be at a TRL 6, which is mature, by the August 2024 CDR. In May 2023, EHP officials said neither the life support system nor the pressure garment system had significantly changed since January 2023—when they were both rated at a TRL 4—and those systems can only be matured through major testing at CDR. Prior to the November 2023 PDR, Axiom will complete the Crew Capability Assessment of the pressure garment system to further mature that technology. For CDR, Axiom will conduct human testing of the life support system in a NASA vacuum chamber facility.

[18] GAO, Technology Readiness Assessment Guide: Best Practices for Evaluating the Readiness of Technology for Use in Acquisition Programs and Projects, [Reissued with revisions on Feb. 11, 2020], GAO-20-48G (Washington, D.C.: Jan. 7, 2020). TRLs are a scale of nine levels used to measure a technology's progress, starting with paper studies of a basic concept (TRL 1) and ending with a technology that has proven itself in actual usage in the product's operational environment (TRL 9). A technology is considered mature when it reaches TRL 6, which is when a model or prototype is demonstrated in a relevant environment.

Axiom plans to produce and manufacture some components itself to reduce supply chain risks. However, NASA officials said that using a different source decreases the TRL for those components and, hence, the TRL will need to be reassessed. This is due to the uncertainty of a new manufacturer rather than a regression of the technology itself.

NASA officials said the components that have design changes are also rated at a lower TRL and will need to be matured for use for the Artemis III mission. To support technology maturation efforts, Axiom personnel are developing multiple test rigs for different components of the life support system.

- *Procure suit components.* Axiom's remaining work to develop and procure suit components risks potential delays, which would compress Axiom's window to less than 2 years for delivering suits to NASA by September 2025. Prior to the Artemis III mission, the space suits need to be delivered so they can be integrated into the human landing system. Some of the parts that support critical systems for the space suit—including the life support system—have long lead times and could potentially take 12-18 months to procure from vendors. For example, Axiom plans to outsource certain parts, such as the oxygen regulators, because there are very few companies that make them to the standard needed for space suits. EHP officials said that procurement of components for the life support system is on Axiom's critical path for the schedule as some of these components are highly complex and specialized. EHP officials said that upon delivery, Axiom will have to conduct qualification and acceptance testing of those parts. To alleviate supply chain issues, Axiom now produces 40 of 61 parts in-house which gives it more control over the design and schedule for building the suit.
- *Qualify the suit for flight.* Axiom will qualify the space suits for flight readiness before a crew can use them, but the necessary testing facilities may not be available in time for the Artemis III mission. To certify the space suits for use, Axiom's proposed certification process requires the use of some NASA facilities. Axiom planned to conduct a crew capability assessment at the NASA Johnson Space Center's Active Response Gravity Offload System (ARGOS) facility as part of the PDR in November 2023. However, in August 2023, EHP officials said there was an issue with the equipment at the ARGOS facility. Axiom instead plans to test the space suits at the Partial

Gravity Simulator facility at NASA's Johnson Space Center in October 2023, ahead of PDR. They also said that they are developing the supporting documentation so that the facility is prepared for Axiom to test the space suits. Additionally, Axiom will complete a vacuum test of the space suit at another NASA facility for CDR.

NASA Is Addressing Multiple Cross-program Risks That Affect Human Landing System and Space Suit Development

In addition to managing the remaining contractor work, NASA is also addressing multiple cross-program risks related to integrating the lander and space suits with systems needed for the Artemis III mission. Cross- program risks involve more than one program and require the programs to coordinate on the mitigation steps. These cross-program risks include, among other things, lander software and hardware integration and lunar dust contamination.

- *Lander software and hardware integration.* The HLS program is addressing a cross-program risk related to integrating the system's software and hardware with the Orion program. The HLS system uses software located across multiple hardware systems and subsystems, which makes it difficult to perform end-to-end software testing on flight-like hardware in relevant mission environments. HLS risk management officials said the program will need emulators and simulators to conduct HLS-Orion joint verification testing.[19] NASA officials said the Orion and HLS programs have made agreements to share four sets of emulator and simulator hardware and software and agreed to the details and exchange dates for each. However, in July 2023, NASA documentation stated that the HLS development pace does not align with Orion program integration milestones and could hinder the planned December 2025 launch readiness date.

  Integration of software developed for dissimilar hardware platforms, using different operating systems, as well as use and integration of heritage software, can be challenging and prone to

[19] Emulators and simulators are software tools used to test critical systems under conditions and environments often unattainable in the lab or test bench. Simulators provide realistic data as would be experienced in actual flight while an emulator is a statistical approximation of the simulator.

introducing defects.[20] HLS risk documentation states that adequate test facilities and test campaigns are required to avoid late discovery of critical software defects because critical issues are often uncovered when software is integrated and tested with flight hardware. Therefore, without adequate testing, it is possible that critical software defects are missed. This situation could result in cost and schedule effects, or worse, potential loss of mission or crew.

The HLS program and SpaceX have discussed applicable lessons learned from Boeing's Starliner first Uncrewed Flight Test conducted for CCP. This flight test failed to reach orbit due to software issues. The lessons learned included the need to increase the resources dedicated to software insight and oversight. The HLS program incorporated these lessons learned into guidelines on how NASA personnel should consider SpaceX-provided data on software development to ensure defects are identified earlier in development.

- *Lunar dust contamination.* Lunar dust is abrasive and is hazardous to the crew and equipment. For example, inhalation of lunar dust can cause irritation of the respiratory system and contact with the dust can cause irritation of the eyes. NASA's Artemis III program offices, including HLS, EHP, and Orion, are working to resolve a potential risk of lunar dust intruding into hardware and negatively affecting the performance of those systems. If not adequately mitigated early in the design process, it may cost more and take longer to retrofit the hardware to safely address lunar dust contamination.

  Each program is working on its own portion of limiting dust exposure. For example, the HLS program is testing cleaning techniques for the lander's hatches and Axiom is developing tools to clean the space suit following a moonwalk. EHP officials expressed concern about the amount of dust that the space suits will pick up and bring into the HLS. Even if each system adequately meets its dust contamination requirement, there is still a risk that crew could become injured or ill due to the dust. Before the Artemis III mission, Axiom plans to demonstrate that it can limit the amount of dust on the exterior of the suit that is brought into the lander's cabin environment. SpaceX plans to demonstrate that the HLS Starship is capable of

[20] Legacy and heritage software are software products written specifically for one project and then, without prior planning during their initial development, found to be useful for other projects.

limiting lunar dust to a defined, acceptable level to safeguard crew health.

NASA's Moon to Mars Program Office—currently responsible for integrating all of the Artemis III systems—is also addressing performance and safety risks related to the known harm that lunar dust could inflict on the hardware and crew. It is addressing these risks across the HLS, space suits, and Orion programs based on lessons learned from the Apollo mission and subsequent research. Additionally, the HLS program coordinated with other NASA personnel and programs to develop a long-term mitigation plan focused on increasing their knowledge of south pole lunar dust and its properties.

NASA Is Taking Steps to Ensure Systems Are Safe and Meet Its Needs before Launch

NASA plans to take multiple steps to determine whether SpaceX's and Axiom's systems meet its mission needs and are safe for the crew. The HLS and EHP programs will ultimately determine whether the contractors' systems meet contract requirements. Then, NASA will conduct a to-be- decided series of reviews to determine whether the agency is ready for launch based on guidance it is currently developing. Further, NASA's contracts with SpaceX and Axiom grant NASA visibility into many areas of contractor work while the companies complete their significant remaining work.

NASA Will Determine If Contractors' Systems Meet Mission Needs, Including Safety

Before NASA conducts the Artemis III mission, it plans to determine whether SpaceX's and Axiom's systems meet requirements and are safe. NASA's requirements documents include requirements that the human landing system and space suits must meet. We have categorized these requirements into two types: system (includes functional and performance) and interface. System requirements include what functions the system needs to perform to accomplish the objectives and how well the system needs to perform the functions. Interface requirements describe how the HLS Starship interfaces with other systems—such as the Artemis space suits and Orion—so that each system will be able to safely operate with the other, among other things. See table 1 for examples of system and interface requirements.

NASA's system requirements for the HLS Starship and space suits include NASA-approved alternative technical standards that the systems must meet:

- The HLS program allowed SpaceX to use alternative technical standards in three areas as long as NASA determined that the alternative standards met the intent of NASA's technical standards. These areas were safety and mission assurance, health and medical, and engineering technical standards. NASA officials said that after they completed the adjudication process with SpaceX, approximately 50 percent of SpaceX's technical standards were alternative to NASA's.
- Similarly, NASA allowed Axiom to propose alternative or tailored standards to NASA's design and construction, safety, and human health and medical standards in its initial proposal. EHP officials said they accepted nine tailored or alternative standards and determined that none of the changes created additional risk to NASA.

Table 1. Examples of requirements applicable to SpaceX and Axiom, by GAO-identified type

	Human Landing System - SpaceX	Space suit - Axiom
Examples of system requirements	Be capable of operating on the lunar surface for a minimum of 6.5 Earth days In the event of an aborted attempt at a lunar landing, have the capability to conduct a safe return and dock to Orion in near-rectilinear halo orbit	Sustain the life of the crewmember for a minimum of 8 continuous hours of operation independent of vehicle-provided life support functions Provide safe and accurate visual capability and head mobility to perform tasks in both daytime and nighttime conditions
Examples of interface requirements	Provide oxygen to the space suit system at a temperature in the range of 40 to 90 degrees Fahrenheit for space suit system recharge Have a passageway providing for the transfer of crew and cargo to and from Orion	The space suits shall provide an interface for the receiving of oxygen for space suit recharge, in-suit pre-breath, and umbilical operations

Source: GAO analysis of SpaceX and Axiom contracts with the National Aeronautics and Space Administration (NASA). | GAO-24-106256.

NASA officials from both programs reviewed the alternative technical standards early in system development. HLS officials said they completed this work early based on lessons learned from CCP to promote a shared understanding of technical standards and avoid late disagreements that could cause delays. EHP officials said that undergoing this process early in development was consistent with NASA policy and a good engineering practice to avoid taking unknown risks.

The programs and contractors each have a role in determining whether the HLS Starship and space suits meet NASA's requirements. SpaceX and Axiom will conduct activities—such as analysis, test, demonstration, or inspection—to verify that their systems meet NASA's requirements.

The contractors will submit their verification results to NASA for approval or disapproval. For example, initial submissions are due at PDR for Axiom.

While NASA policy allows but does not require design certification reviews, SpaceX and Axiom are contractually required to undergo this review before the Artemis III mission to ensure that their designs comply with system and interface requirements.[21] At these reviews, NASA will review each contractor's design and evidence to ensure that the contractor's system meets all system and interface requirements, among other things.

NASA Is Formulating Plans to Determine Agency Readiness to Launch

In December 2022, NASA's Artemis Campaign Development Division released an implementation plan for Artemis missions that provides a high-level summary of NASA's planned approach to technical and programmatic reviews, among other things.[22] The implementation plan states that after each of the programs' design certification reviews, there will be a design certification review of the integrated architecture requirements for Artemis III. This review will assess whether the five- program integrated system—HLS,

[21] NASA Procedural Requirements (NPR) 7123.1C, NASA Systems Engineering Processes and Requirements (Feb. 14, 2020).

[22] The Artemis Campaign Development Division Implementation Plan was released before NASA announced that it established the Moon to Mars program office. A January 2023 NASA report on Moon to Mars program implementation stated that the program will leverage the work done by the Artemis Campaign Development Division and will update the existing implementation plan to reflect any necessary changes. This plan will generally apply to the Artemis III mission and beyond and the applicable programs.

EHP, Orion, SLS, and EGS—can meet requirements across all applicable configurations and environments, among other things. Then NASA will undergo a to-be-determined certification of flight readiness process to determine if these programs are ready to execute the mission (see fig. 7).

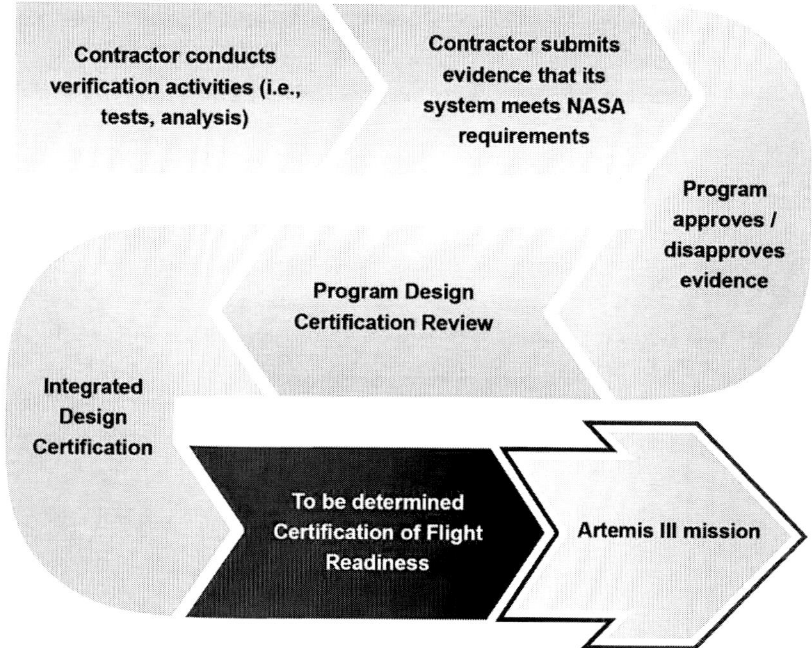

Source: GAO analysis of NASA, SpaceX, and Axiom plans and contract documents, and interviews with NASA officials. | GAO-24-106256.

Figure 7. NASA Processes to Determine Whether a Contractor's System Meets Requirements before Key Reviews to Support Artemis III Launch.

NASA officials said they are currently developing general guidance for implementing certification of flight readiness for future Artemis missions as well as Artemis III mission-specific guidance.

- NASA officials said the general Artemis certification of flight readiness guidance will likely establish a framework for programs to develop their own certification of flight readiness plans and define the reporting structure for Artemis flight readiness, among other things. NASA plans to obtain a formal, integrated human rating certification for the Artemis III mission and associated crewed space system architecture as part of the certification of flight readiness process. NASA policy states that a human-rated system is required for crewed space systems to control hazards, to manage safety risks associated with human spaceflight, and safely recover crew.[23]

- Officials also said the Artemis III mission-specific certification of flight readiness guidance will likely explain the timeline and sequence of multiple flight readiness reviews that will take place to determine readiness to launch.

  The flight readiness review (1) examines tests, demonstrations, analyses, and audits that determine if a system is ready for a safe and successful flight or launch; and (2) ensures all flight and ground hardware, software, personnel, and procedures are operationally ready. Officials said they anticipate that the agency will hold at least eight flight readiness reviews to determine readiness for Artemis III:
  - five program-level reviews (HLS, EHP, Orion, SLS, EGS),
  - two Moon to Mars program-level reviews, and
  - one agency-level review led by the Exploration Systems Development Mission Directorate Associate Administrator, who is responsible for assessing the flight readiness of programs and projects within the mission directorate.

NASA officials said they are applying NASA guidance and best practices from earlier human spaceflight efforts to their Artemis certification of flight readiness plans. These earlier efforts include the Space Shuttle, CCP, and ISS missions. For example, officials said that like ISS missions, the Artemis III certification of flight readiness process will begin around 3 months before launch. However, they said there are limitations with this approach due to the complexity of integrating five systems for the Artemis III mission. There are multiple launches that will take place for the Artemis III mission, which will include the Starship depot and tankers, as well as the HLS Starship and the crew on Orion. Officials said NASA will have to determine the timing and spacing of the reviews for the Artemis III mission, given the number of launches needed to execute the mission.

NASA officials said they plan to release both the general Artemis certification of flight readiness plan and the Artemis III mission supplement

---

[23] NASA Procedural Requirement (NPR) 8705.2C, *Human Rating Requirements for Space Systems* (July 10, 2017).

in 2024. They stated that they plan to release both plans 1 year before launch to give them enough time to be implemented for Artemis III.

NASA's Contracts Grant Visibility into Technical Progress and Safety during Development

NASA included clauses in the contracts with SpaceX and Axiom to gain visibility into contractor efforts under the firm-fixed-price contracts for services. The SpaceX and Axiom contracts grant NASA visibility into contractor efforts throughout development and at key milestones as follows.

Approval authority. Both contracts require the contractors to submit specific data deliverables at key milestones and identify which deliverables must be approved by NASA. For example,

- Both contracts require a system safety assessment report that documents potential safety hazards and methods to control those hazards. The contractors must submit these safety reports for NASA review and approval at multiple milestone reviews to support program and independent safety review panels at NASA.
- Each month, both contractors are also required to submit an integrated master schedule that documents planned work, and resources necessary to accomplish that work. Axiom's monthly integrated master schedule submission must include a schedule risk analysis.[24] SpaceX must submit a schedule risk assessment and an integrated master schedule, at the critical design review and design certification review. SpaceX's integrated master schedule does not need to be approved by NASA, but it will be used to plan, manage, and report work required in performance of the contract. In contrast, NASA has a time-limited right to disapprove Axiom's integrated master schedule.

[24] Our schedule assessment guide defines schedule risk analysis as incorporating program schedule risks into a statistical simulation to predict the level of confidence in meeting a program's completion date; to determine the contingency, or reserve of time, needed for a level of confidence; and to identify high-priority risks. This analysis should be performed before a baseline is set. GAO, *GAO Schedule Assessment Guide: Best Practices for Project Schedules*, GAO-16-89G (Washington, D.C.: December 2015).

The HLS program manager said that while the program has yet to experience delays due to needing time to review SpaceX data, program staff recognize that timely review will be critical as SpaceX's development progresses. The program manager said that the program developed a schedule for reviewing SpaceX data to provide its technical opinion on a timely basis.

Table 2. Scope of NASA's insight into SpaceX and Axiom activities

SpaceX contract	Axiom contract
Areas of insight include: • Any aspect of the design, development, analysis, testing, schedules, performance metrics, risks, and management processes of the contractor's human landing system and individual vehicles, elements, integrated systems, subsystems, etc. According to NASA, this includes every aspect of the integrated lander, all supporting spacecraft, and if applicable, any Active-Active docking adapter. • Launch vehicles and launch site operations. According to NASA, this includes, but is not limited to, spacecraft-to- launch vehicle integration, spacecraft handling procedures, launch commit criteria, and range safety analysis supporting the launch of the integrated lander elements, supporting spacecraft, and all payloads, including non-NASA payloads or cargo. • Flight and mission operations. According to NASA, this includes preparations, flight plans, rules and procedures, trajectory and mission design, crew and flight control team training, real-time operations, space communications and navigation networks, and any non-NASA services performed by the contractor. • Any other contract performance activities or data identified by NASA that are mission-critical or otherwise related to safety in any manner.	NASA has insight into Axiom, subcontractor, and partner entities' efforts that could affect Artemis requirements, interfaces, integration, operations, and crew safety. These include: • Design • Development • Manufacturing • Management • Mission integration • Vehicle integration • Operations • Medical and health • Training and certification • Hardware/software testing • NASA has access to all Axiom activities associated with Artemis interface compatibility and safety certification. NASA has access into any Axiom-initiated space suit changes across its commercial endeavors or any changes that may affect NASA missions.

Source: GAO analysis of SpaceX and Axiom contracts with the National Aeronautics and Space Administration (NASA). | GAO-24-106256.

Insight. The SpaceX and Axiom contracts grant NASA insight or access into many areas of contractor efforts, including any changes that could affect the mission or crew safety. SpaceX and Axiom provide NASA staff insight through the use of recurring meetings, electronic systems, and access to their facilities. Both contracts also grant NASA insight into certain aspects of SpaceX's and Axiom's commercial variants of the Starship and space suits, respectively. For example, HLS officials said that insight allowed them to observe SpaceX's activities leading up to the Orbital Flight Test, which flew a commercial Starship variant. Both contracts include a government insight clause that lays out the scope of NASA's insight. See table 2.

NASA officials said that the purpose of ensuring insight into contractor efforts is to verify technical information, which should help ensure that formal milestone reviews and deliverable submissions are successful. For example:

- HLS program officials said they were able to use insight clauses and data deliverables to gain visibility into how SpaceX's critical technologies were maturing, even though SpaceX is not required to provide TRL information. Further, they said that information exchanged through insight opportunities helps SpaceX focus its own design and documentation efforts, which should result in higher-quality deliverables for formal reviews.
- EHP officials said that they included insight clauses in Axiom's contract to ensure that NASA has extensive insight into the contractor's design and interfaces, including changes to NASA's government reference design. EHP officials said that insight into contractor activities will be used between formal milestones to maintain awareness of emerging issues on a timely basis. They believe this awareness will reduce the likelihood that Axiom's milestone reviews are delayed or put NASA in a position to rush the review.

As NASA and its contractors are learning to implement the insight clauses, NASA officials said they were cognizant that insight activities could pose a schedule risk to its contractors. For example:

- The HLS program manager said that conducting insight on a schedule-driven, firm-fixed-price contract was a culture shift for NASA staff who were used to conducting oversight on cost-reimbursement contracts.[25] By default, the HLS program is using its access granted by the insight clauses to primarily monitor SpaceX's efforts, but it also established processes to change the depth of its insight. For example, the HLS program could conduct independent analyses to corroborate its understanding of SpaceX's progress. While the program allows for insight levels to be changed based on an ongoing assessment of risks, any change in the level of insight must be approved by the HLS program manager. The program established this process in recognition that deeper insight levels would require additional resources from (1) the HLS program to conduct deeper insight, and (2) SpaceX to address NASA's insight results. HLS officials said that areas in which the program has deepened its insight to date include system safety, propulsion, software, flight mechanics, landing stability, and contingencies.

According to both HLS officials and SpaceX representatives, SpaceX gave NASA insight beyond what the contract requires by sharing information about its work under other programs as well as its Starship development. The program reported that its insight into SpaceX's early Starship development has been beneficial for NASA and SpaceX as they solve problems and reduce risk. SpaceX representatives explained that they were sharing information with NASA because these early efforts will inform what is needed for the HLS Starship. They also said they carried over "crew office hours" that were used with CCP, where NASA astronauts can have detailed discussions with the SpaceX team on topics such as mission operations.

- EHP officials said that Axiom's contract gives NASA the ability to dig deep into any area that poses risk and there is no barrier for NASA to seek access. Therefore, officials stated that insight activities represent schedule risk to Axiom because the contractor does not know how much time or how many staff will be needed to address or accommodate NASA's insight activities. EHP officials also said that they have already learned that insight processes were bottlenecked by

[25] Cost-reimbursement types of contracts provide for payment of allowable incurred costs, to the extent prescribed in the contract. These contracts establish an estimate of total cost for the purpose of obligating funds and establishing a ceiling that the contractor may not exceed (except at its own risk) without the approval of the contracting officer. FAR 16.301.1. Cost-reimbursement contracts require the government to maintain extensive visibility into a contractor's technical progress and financial performance. FAR 16.301- 3(a)(3)(4).

administrative processes, such as how quickly they were able to share information, rather than by technical work.

It is too soon to determine how well the HLS and EHP programs will balance their insight into the contractors' efforts with the schedule pressure to conduct the Artemis III mission as planned in 2025. We previously reported that CCP's insight into contractor efforts took more time than the program or contractors anticipated.[26] More recently, an Aerospace Safety Advisory Panel report stated that NASA continues to have a strong safety culture.[27] Further, in May 2023, the panel chair told us that she does not have any concerns about NASA losing focus on safety, even as pressure to move faster to meet the Artemis III date increases.

Collaboration. NASA included clauses related to the use of government resources in both contracts to formally define the availability of NASA expertise to its contractors. However, the contracts differ in how collaboration relates to approval authority and insight. The Axiom contract states that collaboration is the highest form of insight, which EHP officials explained to mean that knowledge gained through collaboration can be used to understand the system and its risks. In contrast, the HLS contract states that collaboration and insight are distinct and governed by uniquely applicable terms and conditions. Both contracts state that NASA has the sole authority to determine the amount, type, or duration of support it provides, and the support remains under the supervisory control of NASA.[28] HLS and EHP officials said the clauses were mutually beneficial to NASA and the contractors because they allow the contractors to leverage NASA's expertise while NASA experts get an opportunity to do hands-on work.

Since contract award, both companies are leveraging or planning to leverage NASA support. SpaceX can request up to 60 full-time NASA employees or equivalent support contractor personnel, while Axiom can request up to 25 full-time NASA employees during contract performance. HLS officials said it has provided 36 equivalent personnel to SpaceX to support specific tasks, such as preparation for the Orbital Flight Test.

---

[26] GAO-17-137.

[27] The Aerospace Safety Advisory Panel was established by Congress to provide advice and make recommendations to the NASA Administrator on safety matters. NASA, Aerospace Safety Advisory Panel Annual Report 2022 (Washington, D.C.: January 2023).

[28] SpaceX's contract also allows for NASA or the NASA support contractor to have supervisory control of NASA staff during collaboration, as appropriate. Axiom's contract states that NASA staff remains employed by, and under supervisory control of, NASA staff at all times during collaboration.

They said SpaceX also requested collaboration in areas such as micrometeoroid orbital debris, Raptor engine development, characterization of lunar landing sites, and risk assessment of the SpaceX-Starship Pad 39A at Kennedy Space Center. Axiom requested collaboration support in 25 areas such as manufacturing, lighting, and crew training.

Agency Comments

We provided a copy of this report to NASA for review and comment. NASA provided technical comments, which we incorporated as appropriate.

We are sending copies of this report to the appropriate congressional committees and the Administrator of NASA. In addition, this report is available at no charge on the GAO website at http://www.gao.gov.

If you or your staff have any questions about this report, please contact me at (202) 512-4841 or RussellW@gao.gov. Contact points for our Offices of Congressional Relations and Public Affairs may be found on the last page of this report. GAO staff who made major contributions to this report are listed in appendix II.

W. William Russell IV

William Russell
Director, Contracting and National Security Acquisitions

Appendix I: Objectives, Scope, and Methodology

The objectives of our review were to describe (1) the extent to which the National Aeronautics and Space Administration (NASA) has made progress in developing key systems needed to land humans on the moon in 2025, and (2) the steps NASA is taking to ensure that its lunar landing systems contractors are developing systems that meet NASA mission needs and are safe for crew. This is the latest in a series of GAO reports addressing NASA's Artemis enterprise.[29] This report focuses on the human landing system (HLS)

[29] GAO, NASA Lunar Programs: Improved Mission Guidance Needed as Artemis Complexity Grows, GAO-22-105323 (Washington, D.C.: Sept. 8, 2022); NASA Lunar Programs:

initial capability, as well as the Artemis Exploration Extravehicular Activity systems—referred to as space suits.[30]

To determine the extent to which NASA made progress in developing key systems needed to land humans on the moon in 2025, we reviewed lunar landing system programs' plans, contract documentation, and quarterly program status reviews to identify program milestones and critical technology demonstrations. The HLS program, Extravehicular Activities (EVA) and Human Surface Mobility program (EHP), and EVA Development project are overseeing the contractor-led development of these systems. For both the HLS and the space suits, we selected milestones from contract and non-contract sources to track program progress for developing these systems. To select milestones, we identified test flight events and technology demonstrations from contract documents, NASA policies on program life-cycle reviews and project management, and schedule information collected from the HLS and EHP programs. We compared actual and planned schedule milestones for HLS as of June 2022, the earliest date that data were available, to July 2023 data. We compared actual and planned schedule milestones for the EVA Development Project—i.e., the space suit project—as of the contract award in September 2022 and compared them to the latest information available as of September 2023.

To examine the planned development time frames for the HLS program relative to time frames for other major NASA projects, we calculated the number of months from program start to completion for these programs and compared the average number to HLS planned time frames. We used cost and schedule data collected for our prior work assessing selected major NASA projects and reviewed data reliability assessments completed for that work. Based on our review of data reliability assessments for that work, we determined the data are reliable for our purposes as the relevant data, time frames, and analyses were the same. We also calculated major projects' percentage of schedule use for key reviews, which included preliminary

---

Significant Work Remains, Underscoring Challenges to Achieving Moon Landing in 2024, GAO-21-330 (Washington, D.C.: May 26, 2021); NASA Human Space Exploration: Significant Investments in Future Capabilities Require Strengthened Management Oversight, GAO-21-105 (Washington, D.C.: Dec. 15, 2020); and NASA Lunar Programs: Opportunities Exist to Strengthen Analyses and Plans for Moon Landing, GAO-20-68 (Washington, D.C.: Dec. 19, 2019).

[30] In this report, we focused on landing systems needed for Artemis III, specifically the human landing system initial capability and the Artemis space suits. NASA is also pursuing a sustained lunar landing capability and modernized space suits for the ISS from industry. However, we did not include these efforts because they are not included in the Artemis III mission.

design review (PDR) and key decision point (KDP) C for projects that were included in our assessment of major project reports and launched between 2010 and 2022.[31] In addition to the HLS program, we analyzed 29 spaceflight projects in our dataset, and excluded non-spaceflight projects completed within the above time frame.

We determined the scope of remaining development work for both programs to meet the planned Artemis III mission date in December 2025 by reviewing program and contractor schedule milestones, risk charts, plans for mitigating risks, and technology maturation plans. Additionally, we analyzed these documents to determine the current operational and technical risks facing lunar landing system acquisitions and assess the remaining work for the programs and contractors. We interviewed officials from the HLS and EHP programs, as well as SpaceX and Axiom personnel, to understand the status of the lunar landing systems programs.

To determine what steps NASA has taken to ensure the contractors will deliver systems that meet mission needs and ensure crew safety, we assessed HLS and EHP program documentation and SpaceX and Axiom contract and system requirements documentation. We collected and reviewed contracts and associated attachments that outline statements of work and requirements and contractor verification and validation plans, as well as NASA policy on human spaceflight safety. We interviewed HLS and EHP program officials, which included the EVA Development project and the Commercial Crew and Cargo program officials. We also interviewed SpaceX and Axiom personnel. We reviewed SpaceX and Axiom insight implementation plans and interviewed NASA and contractor personnel regarding contract execution. We also reviewed NASA agendas and presentation slides on lessons learned from using a service contract approach, as well as mission integration documentation. We interviewed NASA Artemis Campaign Division officials to understand their certification of flight readiness plans for the Artemis III mission.

We conducted this performance audit from September 2022 to November 2023 in accordance with generally accepted government auditing standards. Those standards require that we plan and perform the audit to obtain sufficient, appropriate evidence to provide a reasonable basis for our findings and

---

[31] GAO, *NASA: Assessments of Major Projects*. GAO-23-106021 (Washington, D.C.: May 31, 2023); and *NASA: Assessments of Major Projects*, GAO-22-105212 (Washington, D.C.: June 23, 2022). When NASA determines that a project has an estimated life-cycle cost of over $250 million, we include that project in our annual review up through launch or the project's end of development.

conclusions based on our audit objectives. We believe that the evidence obtained provides a reasonable basis for our findings and conclusions based on our audit objectives.

Appendix II: GAO Contact and Staff Acknowledgments

GAO Contact

William Russell, (202) 512-4841 or RussellW@gao.gov.

Staff Acknowledgments

In addition to the contact named above, Kristin Van Wychen (Assistant Director); Erin Roosa (Analyst-in-Charge); John Armstrong; Susan Ditto; Lorraine Ettaro; Nathan Hanks; Tonya Humiston; Erin Kennedy; Joy Kim; Meredith Allen Kimmett; Douglas Luo; Edward J. SanFilippo; Sylvia Schatz; Kate Sharkey; Kevin Walsh; and Alyssa Weir made significant contributions to this report.

Related GAO Products

NASA: Assessments of Major Projects. GAO-23-106021. Washington, D.C.: May 31, 2023.

NASA Lunar Programs: Improved Mission Guidance Needed as Artemis Complexity Grows. GAO-22-105323. Washington, D.C.: September 8, 2022.

NASA Lunar Programs: Moon Landing Plans Are Advancing but Challenges Remain. GAO-22-105533. Washington, D.C.: March 1, 2022.

NASA: Lessons from Ongoing Major Projects Could Improve Future Outcomes. GAO-22-105709. Washington, D.C.: February 9, 2022.

NASA Lunar Programs: Significant Work Remains, Underscoring Challenges to Achieving Moon Landing in 2024. GAO-21-330. Washington, D.C.: May 26, 2021.

James Webb Space Telescope: Project Nearing Completion, but Work to Resolve Challenges Continues. GAO-21-406. Washington, D.C.: May 13, 2021.

NASA Human Space Exploration: Significant Investments in Future Capabilities Require Strengthened Management Oversight. GAO-21-105. Washington, D.C.: December 15, 2020.

NASA Commercial Crew Program: Significant Work Remains to Begin Operational Missions to the Space Station. GAO-20-121. Washington, D.C.: January 29, 2020.

NASA Lunar Programs: Opportunities Exist to Strengthen Analyses and Plans for Moon Landing. GAO-20-68. Washington, D.C.: December 19, 2019.

NASA Commercial Crew Program: Schedule Uncertainty Persists for Start of Operational Missions to the International Space Station. GAO-19-504. Washington, D.C.: June 20, 2019.

GAO's Mission

The Government Accountability Office, the audit, evaluation, and investigative arm of Congress, exists to support Congress in meeting its constitutional responsibilities and to help improve the performance and accountability of the federal government for the American people. GAO examines the use of public funds; evaluates federal programs and policies; and provides analyses, recommendations, and other assistance to help Congress make informed oversight, policy, and funding decisions. GAO's commitment to good government is reflected in its core values of accountability, integrity, and reliability.

Connect with GAO

Connect with GAO on Facebook, Flickr, Twitter, and YouTube. Subscribe to our RSS Feeds or Email Updates. Listen to our Podcasts. Visit GAO on the web at https://www.gao.gov.

To Report Fraud, Waste, and Abuse in Federal Programs

Contact FraudNet:

Website: https://www.gao.gov/about/what-gao-does/fraudnet Automated answering system: (800) 424-5454 or (202) 512-7700.

Congressional Relations

A. Nicole Clowers, Managing Director, ClowersA@gao.gov, (202) 512-4400, U.S. Government Accountability Office, 441 G Street NW, Room 7125, Washington, DC 20548.

Public Affairs

Chuck Young, Managing Director, youngc1@gao.gov, (202) 512-4800.

U.S. Government Accountability Office, 441 G Street NW, Room 7149 Washington, DC 20548.

Strategic Planning and External Liaison

Stephen J. Sanford, Managing Director, spel@gao.gov, (202) 512-4707.

U.S. Government Accountability Office, 441 G Street NW, Room 7814, Washington, DC 20548.

Chapter 4

Artemis Programs: NASA Should Document and Communicate Plans to Address Gateway's Mass Risk*

United States Government Accountability Office

Abbreviations

ACR	Architecture Concept Review
CMV	comanifested vehicle
DSL	Deep Space Logistics
ESDMD	Exploration Systems Development Mission Directorate
ESPRIT-RM	European System Providing Refueling, Infrastructure, and Telecommunications Refueler Module
HALO	Habitation and Logistics Outpost
HLS	Human Landing System
I-HAB	International Habitat
JCL	joint cost and schedule confidence level
KDP	key decision point
M2M	Moon to Mars
NASA	National Aeronautics and Space Administration
NRHO	near rectilinear halo orbit

---

* This is an edited, reformatted and augmented version of the United States Government Accountability Office Report to Congressional Committees, Publication No. GAO-24-106878, dated July 2024.

In: Back to the Moon
Editor: Silas Blackwood
ISBN: 979-8-89530-861-5

Orion	Orion Multi-Purpose Crew Vehicle
PPE	Power and Propulsion Element
SAO	Strategy and Architecture Office
SEP	Solar Electric Propulsion
SLS	Space Launch System
SpaceX	Space Exploration Technologies Corporation

Why GAO Did This Study

NASA plans to return astronauts to the moon to make new scientific discoveries, generate economic benefits, and inspire a new generation. To help support crewed lunar landings, NASA plans to use the Gateway as a habitat and safe work environment for astronauts. NASA plans to first use the Gateway to house crew during the Artemis IV lunar landing mission, which NASA is planning to conduct in September 2028. NASA tracks the Gateway program's progress via cost and schedule commitments.

A House Report contains a provision for GAO to continue reviewing NASA's lunar-focused programs. This report focuses on the Gateway program and its NASA-led development projects. It addresses (1) the Gateway program's plans to update the initial capability's cost and schedule analysis; (2) the extent to which the Gateway program made progress with its U.S.-led projects needed for the Artemis IV mission and is addressing project risks; and (3) NASA's process for determining how it will use the Gateway beyond Artemis IV, including for Mars missions.

GAO analyzed NASA documentation and interviewed officials on the Gateway program's cost, schedule, risks, and role in the Artemis architecture.

What GAO Recommends

GAO is making one recommendation, that NASA should ensure that the Gateway program documents and communicates an overall mass management plan before its next program-level review. NASA agreed with GAO's recommendation.

What GAO Found

The National Aeronautics and Space Administration (NASA) plans to build a sustained human lunar presence and ultimately travel to Mars through a series of missions known as Artemis. For Artemis IV, the agency is developing the Gateway—the first space station planned to orbit the moon. NASA committed to launching the Gateway initial capability by December 2027 at a cost of $5.3 billion. The launch will include the first components of the Gateway—the Power and Propulsion Element (PPE) and the Habitation and Logistics Outpost (HALO).

The Gateway program plans to update the analysis it used to inform its cost and schedule commitments at a fall 2024 program-level review. This will help determine the feasibility of the Artemis IV mission date. To reach lunar orbit and ensure all systems work as planned, the PPE and HALO need to launch at least 12 months before the Artemis IV mission, or 3 months earlier than Gateway's current committed date. NASA officials said the program plans to work to an accelerated, to-be-determined date that would provide more schedule flexibility.

Gateway Program Launch Date Options for Artemis IV Mission

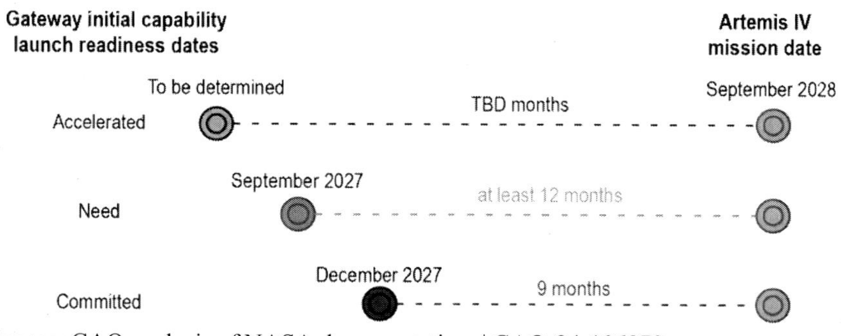

Source: GAO analysis of NASA documentation. | GAO-24-106878.

The Gateway program's projects—including PPE and HALO—made varying degrees of progress over the last year. However, the PPE and HALO projects face several significant challenges. For example, their combined mass is greater than their mass target. Mass is one of many factors that the program considers in its overall design. If they cannot meet their mass target, it may affect their ability to reach the correct lunar orbit. The program has not yet documented an overall mass management plan, which would describe the program's mass reduction approach and priorities for key trade-off decisions. Documenting and communicating this plan will help to ensure that the program and its projects agree on how to address the mass challenge.

NASA held two reviews in 2023 to break down high-level Artemis exploration objectives and goals into the programs, projects, or systems needed to achieve them. So far, NASA has used these reviews to assign roles to the Gateway that align to goals of the earlier Artemis missions, like returning humans to the moon. NASA plans to use upcoming reviews to make key decisions related to Mars missions, which could inform how NASA might use the Gateway in the future.

July 31, 2024

Congressional Committees

The National Aeronautics and Space Administration (NASA) plans to return astronauts to the moon, build a sustainable lunar presence over the next decade, and ultimately travel to Mars through a series of missions known collectively as Artemis. As part of these plans, the agency is developing the Gateway—a small space station planned to orbit the moon. The Gateway will serve as a research platform, a staging point for human and robotic exploration in deep space, and a technology test bed for future Mars exploration. The Gateway is the central aggregation point for the Artemis IV lunar landing mission, currently planned for September 2028. During the mission, the Gateway will house crew before, during, and after the lunar landing. Between fiscal years 2018 and 2029, NASA anticipates spending over $7 billion to build and operate the Gateway.

The Gateway program is composed of three U.S.-led projects: Power and Propulsion Element (PPE), Habitation and Logistics Outpost (HALO), and Deep Space Logistics (DSL). Each of these projects is developing a module that will provide unique capabilities for the Gateway. The PPE and HALO make up the Gateway's initial capability. These two modules together can support a crew on the Gateway, with the PPE providing the power and propulsion and the HALO providing a space for crew to live.

NASA plans to launch the PPE and HALO together prior to the Artemis IV mission, so that the agency is ready to support a crew during the mission. In addition, NASA plans to launch cargo into lunar orbit on a logistics vehicle to support crewed missions to the Gateway, including for the Artemis IV mission.[1] Over time, NASA, and its international partners plan to add modules to the Gateway to support later Artemis missions.

The House Report 117-395 accompanying the Commerce, Justice, Science, and Related Agencies Appropriations bill, 2023 contains a provision for GAO to conduct in-depth reviews of NASA's lunar-focused programs. This report focuses on the Gateway program and its U.S.-led development projects. This report addresses (1) the Gateway program's plans for updating its cost and schedule analysis for the initial capability; (2) the extent to which the Gateway program has made progress with the PPE, HALO, and DSL vehicle needed for the Artemis IV mission in 2028 and is addressing project risks; and (3) NASA's process for determining how it will use the Gateway to support missions beyond Artemis IV, including Mars missions.

To determine the Gateway program's plans to update its cost and schedule analysis for the initial capability, we reviewed NASA and program documentation, including documentation of the Gateway program's initial capability cost and schedule analysis. We interviewed Gateway and Moon to Mars (M2M) program officials to understand the agency's process for developing and approving the cost and schedule baselines for the initial capability and to identify when the program aims to update the cost and schedule analysis. We also reviewed NASA documentation and interviewed program officials to determine when the Gateway program would need to launch the PPE and HALO together for the Gateway to be ready to support the Artemis IV mission. We refer to this date as the need launch readiness date in this report.

To determine the progress of the Gateway program and HALO, PPE, and DSL projects toward supporting the Artemis IV mission and their plans to address project risks, we assessed Gateway program and project documentation and interviewed program and project officials. We reviewed the documentation and interviewed officials to determine how they plan to address technical and design challenges and top risks. We also interviewed M2M program, Gateway program, and HALO project officials to understand how the Gateway fits into the concept of operations for the Artemis IV mission and supports the planned September 2028 mission date. We compared program plans to address technical and design risks against NASA policy and guidance related to program management and systems engineering. We also

[1] The DSL project manages a contract that provides commercial end-to-end services for the delivery of cargo, supplies, and other necessities on logistics vehicles for crew.

compared these plans against our best practices for technology readiness and product development and federal internal control standards.[2]

To understand the agency's processes for determining how it will use the Gateway to support missions beyond Artemis IV, we examined NASA's architecture review process. This process is intended to map high-level M2M objectives to the specific elements that will support science and exploration goals. We reviewed documentation from the first two architecture concept review cycles, M2M and Gateway program requirements documents, and other related documentation. We interviewed M2M program, Strategy and Architecture Office, and Gateway program officials to discuss the architecture concept review process, and the extent to which the Gateway's role in missions beyond Artemis IV has been determined, among other topics.

See appendix I for more information on our objectives, scope, and methodology.

We conducted this performance audit from May 2023 to July 2024 in accordance with generally accepted government auditing standards. Those standards require that we plan and perform the audit to obtain sufficient, appropriate evidence to provide a reasonable basis for our findings and conclusions based on our audit objectives. We believe that the evidence obtained provides a reasonable basis for our findings and conclusions based on our audit objectives.

Background

The Gateway and Its Role in the Artemis IV Mission

The Gateway will help support NASA's long-term lunar exploration goals to create a sustained presence on and around the moon.[3] NASA plans to return astronauts to the moon to make new scientific discoveries, generate economic benefits, and inspire a new generation. NASA first plans to use the Gateway to house crew during the Artemis IV mission.

---

[2] GAO, Standards for Internal Control in the Federal Government, GAO-14-704G (Washington, D.C.: September 2014); Best Practices: Using a Knowledge-Based Approach to Improve Weapon Acquisition, GAO-04-386SP (Washington, D.C.: Jan. 1, 2004); and Best Practices: Capturing Design and Manufacturing Knowledge Early Improves Acquisition Outcomes, GAO-02-701 (Washington, D.C.: July 15, 2002).

[3] NASA, in conjunction with its international partners, also plans to conduct multiple scientific experiments as the Gateway's PPE and HALO transit to the moon.

For Artemis IV, NASA plans to field an initial configuration of the Gateway consisting of three U.S.-developed elements. NASA will then add four additional elements contributed by international partners until it reaches what it calls Gateway's sustained configuration.[4]

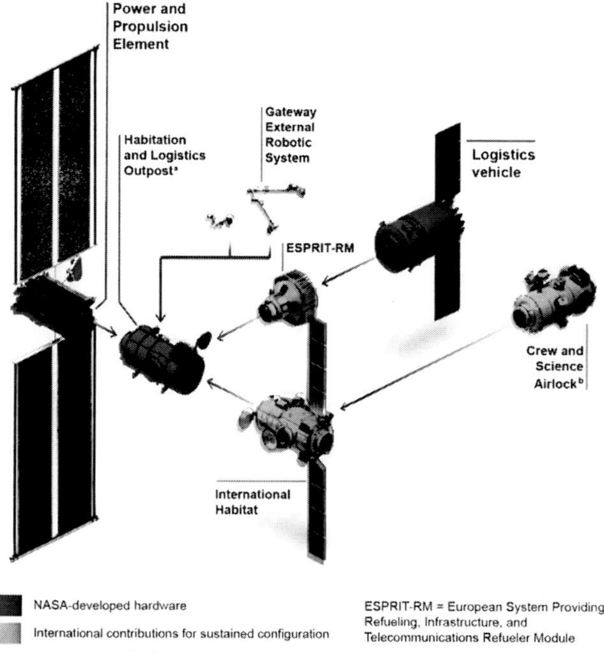

Source: GAO analysos of Gateway program documentation (data); NASA (image). | GAO-24-106878.

[a] The Habitation and Logistics Outpost includes hardware provided by international partners.

[b] The illustration of the Crew and Science Airlock is based on a government reference design.

Figure 1. Illustration of the Gateway Sustained Configuration.

---

[4] NASA plans to add modules developed by international partners to the Gateway during the Artemis IV, V, and VI missions to create the sustained configuration. The sustained configuration adds additional capabilities to support longer crewed missions and additional science operations. For example, NASA signed agreements with the European Space Agency to provide an additional habitation module and a refueler module, the Canadian Space Agency for a robotic arm, and the Mohammed Bin Rashid Space Centre of the United Arab Emirates for a crew and science airlock. NASA also signed an agreement with the Government of Japan to provide life support systems and batteries and another logistics resupply vehicle.

NASA plans to use the Gateway in multiple ways to support Artemis missions. For example, it is to serve as a habitat and safe work environment for astronauts and as a communications relay between the lunar surface and Earth. It will also facilitate lunar landings. Figure 1 shows the planned Gateway sustained configuration.

NASA first plans to use the Gateway in the Artemis IV mission. This mission will be complex because NASA will need to coordinate across seven NASA programs, multiple contractors that support those programs, and international partners to execute the mission. It will also be the first launch of an upgraded version of the Space Launch System rocket.

During the mission, astronauts will arrive at the Gateway on the Orion Multi-purpose Crew Vehicle (Orion), help integrate the International Habitat with the HALO, and conduct a lunar landing.[5] The crew will transfer into a human landing system for transport to the lunar surface and back. After returning to the Gateway, the crew will return to Earth aboard the Orion crew capsule. See figure 2 for more details on key Gateway events in the concept of operations for the Artemis IV mission.

NASA is partnering with industry for the Gateway's U.S.-led projects. NASA awarded contracts to Maxar Space Systems for the PPE, Northrop Grumman for the HALO, and Space Exploration Technologies Corporation (SpaceX) for the logistics vehicle. See table 1 for project descriptions and details on the acquisition strategy the agency is using for each project.

Table 1. U.S.-led Gateway Projects, Project Descriptions, and Acquisition Strategies

Project	Description	Acquisition strategy
Power and Propulsion Element (PPE)	The PPE is to provide the Gateway with power, communications, and the ability to change orbits, among other things.	In May 2019, NASA awarded a firm-fixed price contract to Maxar Space Systems to develop, build, and demonstrate power, propulsion, and communications capabilities. The initial value of the contract was $375 million. As of July 2023, the total value of the contract was over $1 billion. The contract price has grown in large part due to requirements changes and NASA's

[5] The International Habitat will provide additional living space and additional life support systems for crew on the Gateway, which will enable longer crewed missions.

Project	Description	Acquisition strategy
		February 2020 decision to launch the HALO and PPE together.
Habitation and Logistics Outpost (HALO)	The HALO is to provide docking ports for visiting vehicles, space for habitation and storage, and the systems to support crew on board the Gateway.	In June 2020, NASA definitized an undefinitized contract action into a cost-plus-incentive-fee contract with Northrop Grumman Space to develop the preliminary design for the HALO. At that time, the cost of the contract was valued at $187 million. In July 2021, NASA incorporated a firm-fixed-price contract modification to add work for the HALO's manufacturing and integration with PPE, among other things. This modification increased the total value of the contract to nearly $1.3 billion.
Deep Space Logistics (DSL)	The DSL project manages the Gateway Logistics Services contract, which provides commercial end-to-end services to the Gateway for cargo deliveries, supplies, stowage, and trash disposal prior to crew arrival to maximize the length of crew stays on the Gateway.	In March 2020, NASA awarded an initial indefinite delivery, indefinite quantity contract to Space Exploration Technologies Corporation (SpaceX). The contract guarantees the company a minimum of two logistics missions. Each mission is a firm-fixed- price task order off the contract. SpaceX is responsible for building, integrating, and operating the logistics vehicle. Under the contract, NASA may award task orders to other contractors to compete to provide logistics services for future missions. These contractors would provide similar services as SpaceX. The maximum value of the contract for all missions is $7 billion.

Source: GAO analysis of NASA documentation and contracts. | GAO-24-106878.

The PPE project relies on another NASA project—Solar Electric Propulsion (SEP)—to develop, build, and qualify high-power solar electric propulsion thrusters. The SEP project is responsible for working with Aerojet Rocketdyne to build and test two qualification thrusters and three flight thrusters, which the SEP project will provide to the PPE project as government furnished equipment.

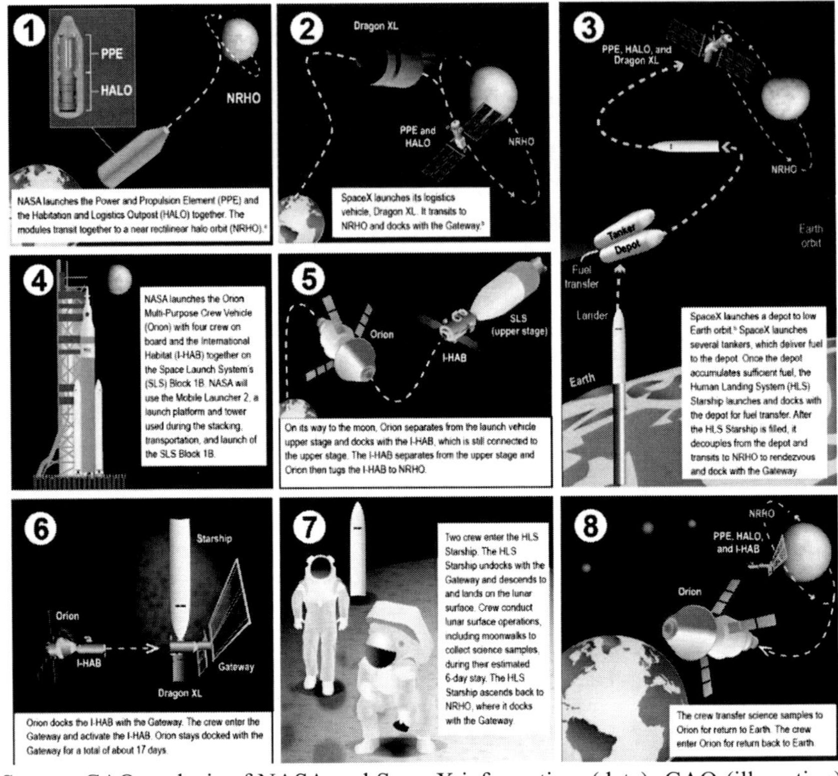

Source: GAO analysis of NASA and SpaceX information. (data); GAO (illusration); GAO illustration of Starship based on SpaceX image. | GAO-24-106878.

[a] Near rectilinear halo orbit is a 1-week lunar orbit balanced between Earth's and the moon's gravity. This orbit enables global lunar access and promotes access to the lunar poles.

[b] NASA has not yet finalized the order of events in steps 2 and 3.

Figure 2. Key Gateway Events in the Artemis IV Mission Concept of Operations.

NASA Acquisition Life Cycle

The Gateway program is NASA's first tightly coupled program, meaning it is composed of multiple projects that work together to complete the program's mission. The acquisition life cycle for a tightly coupled program closely resembles the life cycle for a spaceflight project, which the PPE, HALO, and DSL projects follow. Life cycles for both consist of two phases, formulation and implementation. The formulation phase takes a program or project from concept to preliminary design, and the implementation phase includes building, launching, and operating the system, among other activities. In addition, both programs and projects hold key decision points (KDP) where senior NASA officials approve programs and projects to move to the next phase. For example, tightly coupled programs hold a KDP I review and projects hold a KDP C review before moving from the formulation to the implementation phase.

In December 2023, NASA approved the Gateway initial capability to enter the implementation phase after completing a KDP I review. This KDP I review also served as the KDP C reviews that approved the PPE and HALO projects to enter their implementation phases. The Gateway program plans to hold a separate KDP C review for the DSL project and establish cost and schedule baselines for the development of the first logistics vehicle.

As part of the initial capability review, NASA established the agency baseline commitment for the Gateway initial capability in a December 2023 decision memorandum. The Gateway initial capability's cost baseline is $5.3 billion and the schedule baseline is December 2027.[6] This represents the cost and schedule baselines against which external stakeholders, such as Congress and the Office of Management and Budget, measure the agency's performance. We refer to December 2027 as the baseline launch readiness date in this report. NASA plans to work to an earlier launch readiness date, which we refer to as the accelerated launch readiness date.

To inform the baselines, NASA policy requires each program and project with a life-cycle cost estimated to be greater than $250 million to develop a joint cost and schedule confidence level (JCL). A JCL produces a point- in-time estimate that includes, among other things, all cost and schedule elements from the start of formulation through the end of the system assembly, integration and test, and launch phase. The JCL incorporates and quantifies known risks, assesses the effects of cost and schedule on the estimate as of the time the JCL is conducted, and addresses available annual resources. The results of a JCL indicate the probability of a program or project's success in meeting cost and schedule targets. For example, NASA estimated a 70 percent probability of the Gateway program meeting its cost baseline and the baseline launch readiness date. Typically, the agency approves baselines at a 70 percent probability of the program or project meeting its cost and schedule targets.

[6] The cost baseline includes the costs of the PPE and HALO projects, launch vehicle, and program support for integration and launch.

Throughout the acquisition life cycle, the PPE, HALO, and DSL projects hold technical reviews to assess the maturity of their systems or evaluate the readiness to move to the next phase. For example:

- Near the end of the formulation phase, projects hold a preliminary design review to assess the maturity of their technologies and to determine if their designs are mature enough to proceed with the detailed design activities.
- During the implementation phase, projects hold a critical design review to determine if their designs are stable enough to support proceeding with the final design and fabrication.
- After the critical design review, projects complete a system integration review to evaluate the readiness of the project and associated supporting infrastructure to begin system assembly, integration, and test.

The Gateway program tailored these technical reviews at the program level to assess the maturity of the program across its projects and international contributions. The program calls these synchronization reviews. The synchronization reviews focus on integrated aspects that the PPE and HALO projects do not address through their project-level reviews. The program plans to hold its critical design-informed synchronization review in September 2024.

Gateway program officials said they are still deciding whether they will hold a KDP II review—a review for NASA to assess whether the program has enough margin and an acceptable level of risk to meet its cost baseline and baseline launch readiness date. In our 2019 report on lunar programs, we recommended that the Gateway program hold a KDP II review.[7] NASA concurred with our recommendation, but Gateway program officials said they are still discussing whether to hold any additional KDP reviews for the initial capability. We continue to believe the program should hold this review as the PPE and HALO projects work toward system integration and test.

[7] GAO, *NASA Lunar Programs: Opportunities Exist to Strengthen Analyses and Plans for Moon Landing*, GAO-20-68 (Washington, D.C.: Dec. 19, 2019).

Relevant GAO Reports and Key Past Artemis and Gateway Events

NASA made several changes to the Gateway program since we first began reviewing it in 2019. See figure 3 for a summary of key events and our relevant report findings and recommendations since 2019.

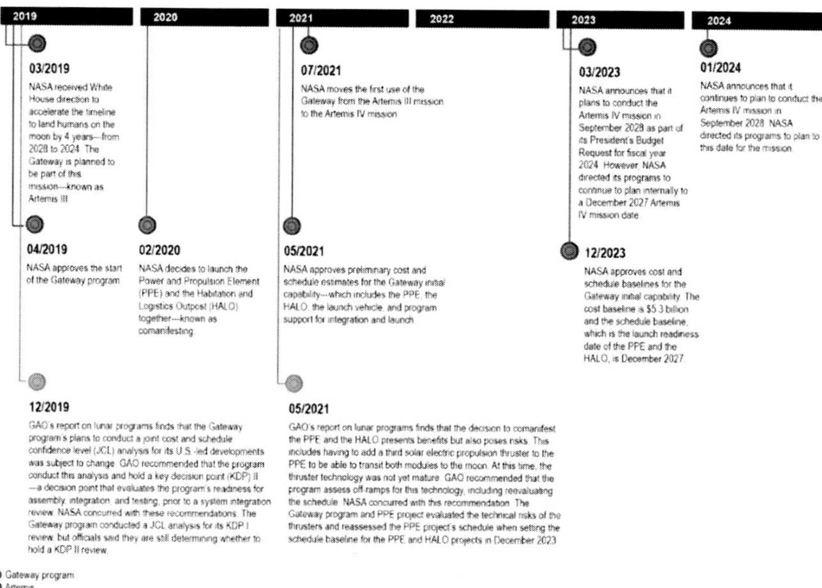

Source: GAO analysis of NASA documentation; GAO (illustration). | GAO-24-106878.

Figure 3. Key Events in the Gateway and Artemis Program's History and Related GAO Report Findings and Recommendations.

NASA acquisition management has been on our high-risk list since 1990.[8] As we noted in our January 2024 testimony on Artemis programs, NASA has made improvements to its acquisition management policies and practices in recent years.[9] However, it still faces challenges in its ability to manage its costliest and most complex programs, such as those that are critical to support the Artemis missions.

---

[8] 8GAO, *High-Risk Series: Efforts Made to Achieve Progress Need to Be Maintained and Expanded to Fully Address All Areas,* GAO-23-106203 (Washington, D.C.: Apr. 20, 2023).

[9] GAO, *NASA Artemis Programs: Lunar Landing Plans are Progressing, but Challenges Remain,* GAO-24-107249 (Washington, D.C.: Jan. 17, 2024).

Establishment of Moon to Mars Program Office and Architecture Concept Review Process

In March 2023, the NASA Administrator established the M2M program office as directed by section 10811 of the National Aeronautics and Space Administration Authorization Act.[10] Under the act, the director of the M2M program office must, among other things, have authority to manage resources, personnel, and contracts to implement the program, and direct and oversee a program-wide systems engineering and integration and integrated risk management function. The office resides within NASA's Exploration Systems Development Mission Directorate (see fig. 4).

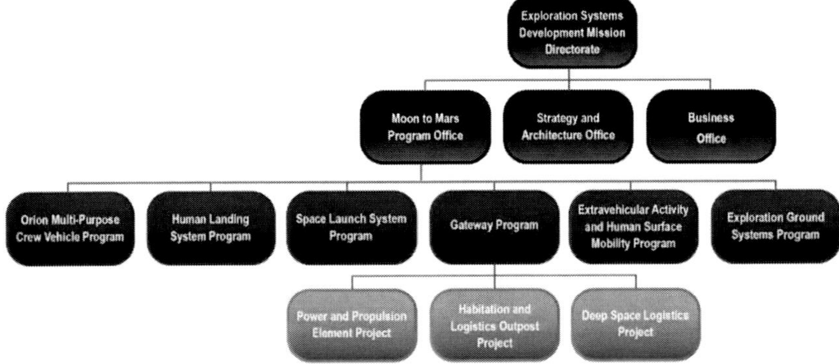

Source: GAO analysis of NASA documentation. | GAO-24-106878.

Figure 4. NASA's Exploration Systems Development Mission Directorate Organizational Chart.

The M2M program office is responsible for supervising the development and operations of the individual M2M programs, including the Gateway. In addition, the program office manages risks for exploration efforts; integrates the design, engineering, operations, and budget formulation for the programs; and oversees Artemis mission preparation, training, operations, and execution.

NASA also created the Strategy and Architecture Office (SAO) within the Exploration Systems Development Mission Directorate in March 2023.

The SAO works alongside the M2M program office and is responsible for defining the agency's architecture for exploration of the moon and Mars based

[10] NASA Authorization Act of 2022, Pub. L. No. 117-167, § 10811 (51 U.S.C. § 20302 note).

on its M2M objectives.[11] This architecture is the agency's high- level unifying structure for its M2M exploration goals. It includes a set of rules, guidelines, and constraints that define the structure and the connections that establish how the individual parts fit and work together. The SAO integrates stakeholder input into the architecture and guides new programs through the pre-formulation phase. Once NASA initiates these new programs, the M2M program office is responsible for their continued development through the remainder of the acquisition life cycle.

In 2022, NASA established the agency's Architecture Concept Review process—the process by which the agency plans to map high-level M2M objectives to the specific elements that will support science and exploration goals.[12] The process centers on an annual study cycle— called the strategic analysis cycle—to continually update and refine the architecture, incorporating feedback from stakeholders from within NASA and across industry, academia, and international partners. The strategic analysis cycle informs architecture decisions by identifying technology gaps and needed capabilities. Each analysis cycle culminates in an annual Architecture Concept Review—a review that brings together NASA leadership to refine the existing architecture and strategies.

After the Architecture Concept Review, the SAO, in coordination with those involved in the review, releases an updated Architecture Definition Document. NASA released the initial Moon-to-Mars Architecture Definition Document in April 2023, and an updated version in January 2024.[13] The primary purpose of the document is to map high-level objectives to specific functions for programs and projects. NASA plans for future revisions to continue to document the mapping of objectives for lunar and Mars missions to existing and new flight programs, projects, and systems.

---

[11] NASA, NASA's Moon to Mars Strategy and Objectives Development: A blueprint for sustained human presence and exploration throughout the Solar System, NP-2023-03-3115-HQ (2023).

[12] See appendix II for more details on this process, including for more information about NASA's annual strategic analysis cycles and architecture concept reviews.

[13] NASA, Exploration Systems Development Mission Directorate, *Moon-to-Mars Architecture Definition Document* (ESDMD-001), NASA/TP – 20230002706 (Washington, D.C.: April 2023); Exploration Systems Development Mission Directorate, *2023 Moon to Mars Architecture Definition Document (ESDMD-001) Revision A,* NASA/TP- 20230017458 (Washington, D.C.: January 2024).

Gateway Program's Cost and Schedule Analysis Update Will Help Inform Artemis IV Launch Date Feasibility

The Gateway program plans to update its JCL analysis, in accordance with NASA policy, for its critical design-informed synchronization review that is currently scheduled for September 2024. This cost and schedule analysis will help officials assess the feasibility of the planned September 2028 Artemis IV mission date. To support this mission, the initial capability must launch 3 months before its December 2027 schedule baseline. As a result, NASA officials plan to set an accelerated date to drive contractor performance and create schedule margin.

The Gateway Program Plans to Update Its Cost and Schedule Analysis

The Gateway program plans to update the JCL analysis for the initial capability at its critical design-informed synchronization review, which is planned for September 2024. This review helps the program determine whether the Gateway's overall design performs as expected and is stable enough to support proceeding with the final design and fabrication. The program plans to hold this review after both the HALO and PPE projects have completed critical design review but prior to them completing integration and test at the project level. The updated JCL will incorporate new risks and assess the effects of cost and schedule on the baselines since the program last conducted the JCL. NASA's policy for management of space flight programs and projects requires projects with a life-cycle cost estimate of over $1 billion, such as the PPE and HALO projects, to update the JCL analysis at critical design review.[14] The policy also requires the program to communicate the updated analysis results to agency senior leaders.

An updated JCL analysis would reflect the Gateway program's current costs, schedule, and risks. Since the program's KDP I review, the HALO and PPE projects have been revising their schedules to accommodate prior milestone delays and a large contract modification, respectively. They also plan to finalize additional contract modifications, although PPE and HALO

[14] At critical design review, programs and projects assess whether they are still on track to meet their cost and schedule baselines.

project officials said they will not know the effects on cost and schedule until the modifications are finalized in mid- to late-2024.

In addition, the updated JCL would reflect the program's dynamic risk posture. For example, in the 2 months between when the program completed its JCL analysis in March 2023 and the program's standing review board review in May 2023, 16 risks changed.[15] The board recommended integrating these changes into the program's JCL, and the program incorporated most of them.

We found that since the program finalized its JCL analysis in May 2023, with the recommended adjustments from its standing review board, the program and its projects have mitigated some risks. However, other risks have grown worse and new ones have emerged. For example:

- The HALO project took steps to mitigate a risk related to the HALO's ability to reduce its heat and control humidity inside the module. The review board was concerned about the HALO's heat management capability at the KDP I review, but the project has since determined a mitigation path that lowered the likelihood that this risk would occur. To help lower the HALO's temperature, the project plans to add software that will turn off non-critical equipment and reduce the module's heat when needed at specific times.
- A program-wide risk concerning the Gateway's communication network has worsened and resulted in new risks emerging, including related risks for Gateway's projects. The network facilitates communication throughout the Gateway. The program discovered several defects on a network chip—which affects multiple Gateway components, including the HALO's flight computer and power distribution system—during testing that could affect the network's functionality, reliability, and performance. For example, these defects could lead the flight computers to unexpectedly restart. If the network is not functioning properly, it could result in loss of control of the Gateway. Program officials are also concerned that they might identify more defects with the communication network, based on the number found already.

  To address this risk, the program formed a study team in October 2023 to determine the root cause of these defects, estimate the

[15] The standing review board is a group of independent reviewers who evaluate the program's technical and programmatic approach, progress, and risk posture.

likelihood of discovering new defects, and recommend how to address them with minimal schedule delays. Program officials said they are working with the hardware contractor to ensure they can incorporate chip updates into the network while the Gateway is on orbit. Officials said they will not know the exact effects on schedule until the program's study team completes its findings in fall 2024.

In addition, by the planned September 2024 synchronization review, the projects will have more information about the timing of key hardware deliveries that affect their integration and test schedules. For example, after the Gateway program conducted its JCL for the KDP I review, the SEP project delayed delivery of three advanced solar electric propulsion flight thrusters to the PPE project for integration. The SEP project, which is managing the assembly of these flight thrusters for the PPE, redesigned its thruster harnesses at the PPE project's direction to fix compatibility issues with a heritage spacecraft component.[16] As of April 2024, the SEP project estimated about a 10-month delay, due to these PPE project-driven changes, to delivering the flight thrusters to the PPE project. However, project officials said they are working with the contractor to streamline their schedule to meet the date by which the PPE project needs the flight thrusters.

The updated JCL analysis to support the September 2024 critical design-informed synchronization review presents program managers with the opportunity to decide whether they need to add more time for uncertainty in their schedules for integration and test activities. Uncertainty accounts for situations in which the program is unable to accurately predict the outcome of a future event. As part of the KDP I review for Gateway, the program's standing review board noted that the program should include more time in its schedule for the PPE's assembly and integration and for integrating the PPE and the HALO together. Further, the program's planned JCL update will help decision-makers determine whether the program has adequate cost and schedule reserves as the HALO and PPE projects enter integration and test, the riskiest development phase.

Updating the Gateway program's JCL analysis will ultimately help NASA determine the feasibility of its September 2028 Artemis IV mission date. M2M program officials also said they are considering conducting a schedule risk

[16] Harnesses are the groupings of wire or cable that transmit signals and electrical power.

analysis for the Artemis IV mission.[17] The program's updated JCL analysis is expected to provide key information for this analysis, such as the current likelihood of activity durations, risks, and opportunities related to the Gateway program.

Gateway Program Is Targeting an Accelerated Launch Date to Align with Planned Artemis IV Mission Date

The Gateway program's schedule baseline does not align with the September 2028 Artemis IV mission date, but the program has a plan to support the mission. There are three key launch readiness dates that NASA is tracking:

- *December 2027 baseline launch readiness date.* This is the program's schedule baseline for the initial capability. This date is 3 months later than needed to support the planned Artemis IV mission date.
- *September 2027 need launch readiness date.* To support the planned September 2028 Artemis IV mission date, program officials said NASA will need to launch the PPE and HALO at least 12 months before the Artemis IV mission, or by September 2027. This is to allow time for the Gateway's initial capability to transit to near rectilinear halo orbit, and for the program to ensure all systems work post-launch and check the orbit's stability before vehicles dock with the HALO. The Gateway program's 2023 JCL results showed that the probability of launching the PPE and HALO in September 2027 is about 50 percent at a slightly lower funding level than the program's cost baseline. Gateway program officials also noted that they continue to mature their performance models and mission operations plans, and they could need an additional 2 to 4 months for transit and to check the PPE and HALO systems in orbit.
- *To be determined accelerated launch readiness date.* Gateway program officials said they recognize the baseline launch readiness date is 3 months after the PPE and HALO need to launch to support the Artemis IV mission. As a result, program officials said they plan

[17] A schedule risk analysis uses statistical techniques to predict the likelihood of a program's completion date, or in this case, the baseline launch readiness date. A JCL analysis uses statistical techniques to predict the likelihood of a program meeting its cost and schedule targets, or in this case, the cost baseline and baseline launch readiness date.

to work to an accelerated date that is earlier than both the need and baseline launch readiness dates. In May 2024, program officials said they anticipated determining the accelerated launch readiness date in summer 2024 after receiving input from others within NASA and international partners.

Without the Gateway's initial capability, the Artemis IV mission cannot proceed as NASA currently envisions. See figure 5 for a comparison of the Gateway initial capability launch readiness dates and the Artemis IV mission date.

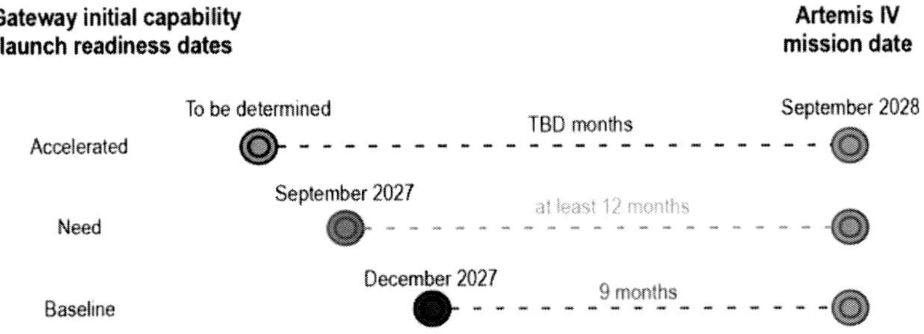

Source: GAO analysis of NASA documentation. | GAO-24-106878.

Figure 5. Gateway Program Launch Date Options for Artemis IV Mission (as of May 2024).

Prior to approving the cost and schedule baseline for the Gateway initial capability, NASA reassessed the feasibility of the Artemis II, III, and IV mission dates. After conducting this assessment, M2M program officials said they took a similar approach with setting baselines for two other Artemis programs. M2M program officials said these programs are all working to dates earlier than their schedule baselines.

An M2M program official said they directed the Gateway and other programs to work to launch readiness dates earlier than their baseline schedules to ensure NASA meets its commitments with Congress and the Office of Management and Budget. The official said programs work with the M2M program, other NASA offices, and their contractors to determine an accelerated but feasible schedule. A feasible schedule keeps contractors working to the earliest date possible without adding risk by asking them to work to an overly optimistic schedule. According to the official, working to an accelerated date helps programs drive contractor performance, create schedule margin, and reduce the risk to the baseline launch readiness date.[18] Gateway program officials said completing work on firm-fixed-price contracts as soon as possible is also in the contractors' best interest as they assume financial responsibility of any additional costs caused by delays.

Given the changes to the PPE and HALO project schedules, the program's dynamic risk posture, and the fact that the initial capability baseline is later than the need date for the Artemis IV mission, it is important that the Gateway program execute its plan to update the JCL analysis at its next synchronization review. As discussed above, the program's planned JCL update in September 2024 will incorporate changes to risks, schedules, and costs since the last JCL and help assess whether the program can support the Artemis IV mission date.

Gateway Program Made Some Progress, but Has Significant Technical Challenges to Overcome Before the Artemis IV Mission

The Gateway program's three projects—PPE, HALO, and DSL—made varying degrees of progress in 2023 and early 2024. While later than initially planned, the PPE and HALO projects entered the final design and fabrication phase and completed their critical design reviews.[19]

The PPE and HALO projects also completed some key tasks that we previously raised as challenges.[20] For example, the PPE project reduced some risk related to its technology readiness by maturing the high-power SEP thrusters. NASA plans to demonstrate the use of these thrusters in deep space on the PPE. In July 2023, the SEP project completed acceptance testing on the first of two qualification model thrusters to mature the technology. This testing included limited vibration, thermal- cycling, and performance testing on the thruster. Similarly, the HALO project completed fabrication of its primary structure. Project officials said they began conducting a key risk mitigation test on the primary structure in May 2024 and plan to complete the test in late

[18] Margin, or schedule reserve, accounts for known risks and uncertainty in the schedule.

[19] When NASA established a preliminary cost and schedule estimate for the Gateway initial capability, the program estimated that the HALO project would complete its critical design review in March 2022 and the PPE project in May 2022. The HALO project delayed its review to allow for time to address design issues and to further mature key subsystems. The PPE project delayed its review to incorporate a high volume of requirements changes into its design.

[20] GAO, *NASA: Assessments of Major Projects,* GAO-23-106021 (Washington, D.C.: May 31, 2023).

June 2024. The goal of the test is to ensure that the structure can withstand the force required to be launched into and operate in space after having to make several welding repairs.

The DSL project progressed into the concept and technology development phase. It is not as far along in its acquisition life cycle as the PPE and HALO projects because the project authorized the start of work on its contract for the design and development of its first logistics vehicle later than planned. NASA awarded SpaceX a contract in March 2020 to develop logistics vehicles to support Artemis missions. However, NASA did not modify its contract with SpaceX to proceed with work to develop and build its first logistics vehicle until November 2023. NASA officials previously told us they delayed ordering the work due to funding constraints and other NASA priorities.[21]

While the PPE and HALO projects made progress toward finalizing their designs, our review of program documentation found that they have several significant technical and design-related challenges to overcome. These include maturing outstanding critical technologies and designs, controlling the Gateway in lunar orbit, and reducing the mass of the combined PPE and HALO for launch and transit to lunar orbit. In addition, M2M-level risks and pending decisions may also affect the Gateway program and its plans.

Delays to maturing technologies and designs. While proceeding through preliminary and critical design reviews, the PPE and HALO projects did not always meet our best practices for technology readiness and design stability.[22] Meeting these best practices can help projects to minimize risk in future phases of development, including limiting future design changes that could result in cost growth and schedule delays. See table 2 for a description of our best practices, the extent to which the projects met them, and project plans to mature outstanding technologies and designs.

Table 2. Extent to which PPE and HALO projects met GAO best practices for technology readiness and design stability as of April 2024

Project	GAO best practice to mature technologies to technology readiness level 6 by preliminary design review[a]	GAO best practice to release 90 percent of design drawings by critical design review[b]
Power and Propulsion Element (PPE)	The PPE project did not mature any of its nine critical technologies at its preliminary design review in 2021.	The PPE project reported that it released over 90 percent of its design drawings at its critical design review

[21] GAO, *NASA: Assessments of Major Projects*, GAO-22-105212 (Washington, D.C.: June 23, 2022).

[22] GAO-04-386SP; and GAO-02-701.

Project	GAO best practice to mature technologies to technology readiness level 6 by preliminary design review[a]	GAO best practice to release 90 percent of design drawings by critical design review[b]
	Since then, the project matured seven of the technologies. The two remaining technologies relate to how the PPE operates its solar electric propulsion thrusters and provides connectivity between the PPE, Gateway, and visiting vehicles. According to project officials, the remaining two technologies will not be mature until after the critical design review, which the project held in March 2024. Project officials said that the project's prime contractor has a different view on the timing of maturing technologies. The officials said they do not view either of the remaining two technologies as a major risk.	in March 2024. However, when a project has immature technologies, it increases the risk of project officials approving a design that is less likely to remain stable.
Habitation and Logistics Outpost (HALO)	The HALO project matured all of its critical technologies by its preliminary design review in 2021.	The HALO project reported that it released 41 percent of its design drawings at its critical design review in 2023. Project officials attributed the low drawing counts primarily to the lower level of maturity of the HALO's environment control and life support subsystem. HALO project officials said that they could mature this subsystem's design later because it sits on a rack or sled device and the contractor can install the entire subsystem at once. As a result, they have primarily focused on maturing the design of the interfaces between the subsystem and the HALO module leading up to a review later in 2024, specifically on this subsystem. Officials said they plan to release additional drawings by the end of 2024, which would get them closer to 90 percent released.

Source: GAO analysis of Gateway program documentation and interviews with officials. | GAO-24-106878.

[a] GAO considers a technology mature for space systems when it reaches a technology readiness level 6, which includes demonstrating a representative prototype of the technology in a relevant environment that simulates the harsh conditions of space. NASA's systems engineering policies align with GAO's technology maturity best practice.

[b] GAO considers engineering drawings a good measure of the demonstrated stability of a product's design because the drawings represent the language used by engineers to communicate to the

manufacturers the details of a new product design—what it looks like, how its components interface, how it functions, how to build it, and what critical materials and processes are required to fabricate and test it. Once the contractor finalizes the design of a product, the drawing is releasable.

Gateway stack controllability. The Gateway program is tracking a risk related to the PPE's ability to keep the Gateway integrated stack in the right orbit and pointing in the right direction when large, heavier vehicles are docked with the Gateway. The integrated stack includes Gateway components like the HALO and other docked spacecraft like lunar landers. Losing precise control of the Gateway integrated stack could result in degradation of performance. For example, if the Gateway is not pointed in the right direction, it could affect communications or the ability for visiting vehicles to successfully dock with the Gateway.

Gateway program officials told us their analysis indicates that there are certain operational scenarios, such as when the lunar lander Starship docks with the Gateway, in which the PPE may not be able to maintain control of the integrated stack. Gateway program officials said that the PPE is meeting the performance requirements for stack controllability that NASA set for it. However, those requirements do not account for the mass of some visiting vehicles that plan to dock with the Gateway. As a result, when these larger than anticipated visiting vehicles dock with the Gateway, the integrated stack may be outside of these controllability parameters (e.g., larger in volume or mass). For example, program officials estimate that the mass of the lunar lander Starship is approximately 18 times greater than the value NASA used to develop the PPE's controllability parameters.

Gateway program officials are conducting additional analyses and studying two main ways to mitigate this risk prior to the program's September 2024 critical design-informed synchronization review. The first way is to have visiting vehicles, such as a logistics vehicle, share some control with the PPE when docked with the Gateway by firing their thrusters for a period, or to require docked visiting vehicle with a mass greater than these original parameters, such as Starship, to control the integrated stack when docked with the Gateway. The second way is making changes to the control algorithms for the PPE to improve control throughout the entire docking process.[23] This includes improving how the program selects different thrusters to fire and to optimize fuel use based on the visiting vehicle that is docking with the Gateway. If neither of these options mitigate the risk, then NASA plans to

---

[23] An algorithm is a set of rules that a computer follows to compute an outcome.

either change the PPE's requirements or add requirements for visiting vehicles. According to NASA's system engineering guidance, late requirements and design changes can lead to cost growth and schedule delays.[24]

Comanifested vehicle (CMV) mass. The HALO and PPE projects are both exceeding their mass allocations as part of the CMV. Prior to launch, NASA plans to integrate the PPE and the HALO together on the ground, which creates the CMV. NASA considers mass a leading indicator—a measure that predicts future performance—and a key measure of design maturity and stability.[25] The Gateway program monitors the overall CMV and each project's mass estimates compared to their allocations to the CMV. If the projects cannot reduce mass to within their allocations, it could affect the CMV's ability to reach the correct lunar orbit. In addition, if the projects need to implement late design changes to reduce their mass, it could result in cost growth or schedule delays. As noted above, NASA's systems engineering guidance indicates that, in general, the later projects make design changes, the larger their negative effect on cost and schedule. The projects have options other than making design changes to reduce mass. These include (1) removing components and reducing capabilities, which would affect performance, (2) carrying components up on a logistics vehicle and having crew install them on-orbit, or (3) narrowing launch windows to those that use less fuel for transit.

The HALO's mass is the primary driver of the CMV exceeding its mass allocation. The HALO's mass increased last year as the project matured its design for internal structures and started receiving more accurate estimates of hardware weights or the hardware itself. In addition, project officials said the HALO's mass increased by 602 kilograms because the project's contractor used an incorrect estimation method to calculate wire harness mass.

See figure 6 for the CMV mass, how much over the mass allocations each project is, and the program's options to address the mass issue.

---

[24] NASA, *NASA Systems Engineering Handbook,* NASA/SP-2016-6105 Rev 2 (Washington, D.C.: 2016).

[25] NASA, *NASA Common Leading Indicators Detailed Reference Guide* (January 2021). Mass is a measurement of how much matter is in an object. It is related to an object's weight, which is mathematically equal to mass multiplied by acceleration due to gravity. When an object goes into space, its weight changes with gravity, but its mass stays the same.

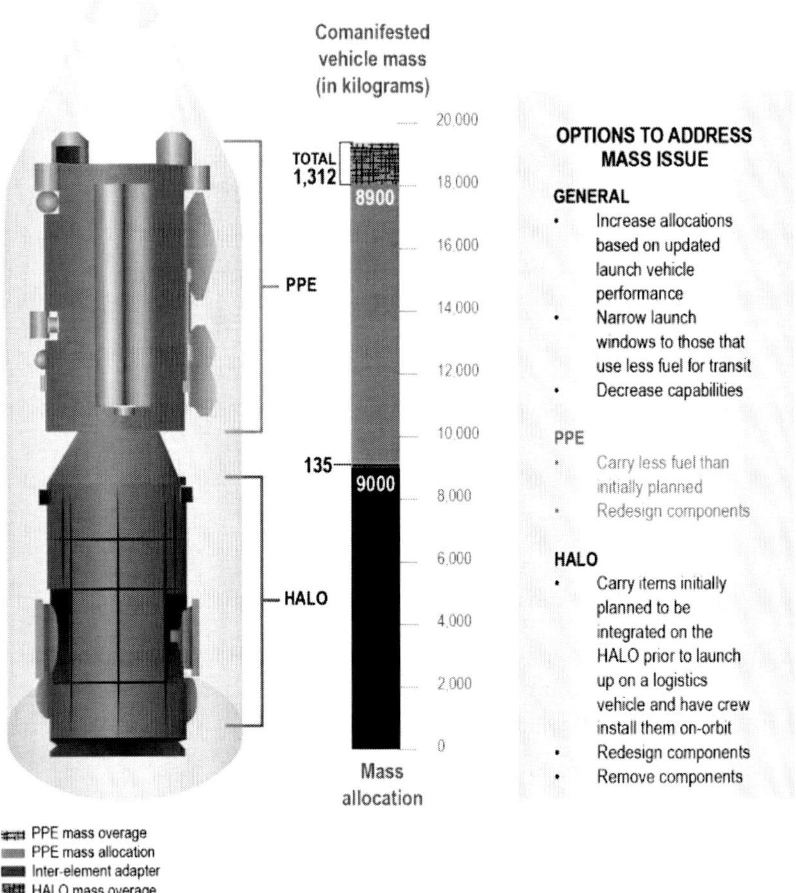

Source: Gatway program documentation and interviews with program and project oficials (data), GAO illustration based on NASA image. | GAO-24-106878.

Figure 6. Comanifested Vehicle Mass and Options to Address Habitation and Logistics Outpost (HALO) and Power and Propulsion Element (PPE) Mass Growth Over Allocations.

NASA's system engineering policy states that one of the criteria for projects to successfully complete the critical design review is to be within their mass allocation with some margin available.[26] The PPE and HALO projects did not meet this criterion at their reviews. Subsequently, the HALO project

[26] NASA, *NASA Systems Engineering Processes and Requirements,* Procedural Requirements 7123.1C (Feb. 14, 2020).

and its contractor began to identify solutions to the mass overage, which will inform a HALO project mass reduction plan. Gateway program officials said that if the projects are not meeting their mass allocation by the program-level critical design-informed synchronization review (scheduled for September 2024), they expect that the projects would submit waivers, or requests to deviate from the requirements, for review at that time.

Gateway program officials said that decisions on how to reduce the mass of the CMV are based on a variety of factors that affect the Gateway's overall mission design and performance. These factors include the predicted mass of HALO and PPE components, power needs, propellant on board, launch vehicle performance, launch windows and trajectories, and the amount of time the CMV needs to transit to lunar orbit. Part of making mass decisions also includes determining how much mass to allocate for HALO components on the logistics vehicle for the Artemis IV mission.

Gateway program officials said they will make decisions on what the first logistics vehicle will carry as they get closer to the 2028 mission. The HALO project identified over 300 kilograms of components to potentially deliver via a logistics vehicle to the Gateway for the crew to install on- orbit. However, the Gateway program will have to balance a decision about whether to carry HALO components on the logistics vehicle with other mission needs. For example, the International Habitat, which the European Space Agency is contributing, is also exceeding its launch mass and may need the logistics vehicle to carry up items to reduce its mass. In addition, the logistics vehicle will also need to carry up cargo; consumables, such as water and food that the crew will need for the mission; and other items.

The Gateway program oversees the efforts to reduce the CMV mass, including determining how to reduce it and what the logistics vehicle will carry. Program officials said they have verbally discussed an approach to reducing mass with the projects, but the program office has not yet documented an overall plan to manage the PPE and HALO mass reduction efforts. Standards for Internal Control in the Federal *Government* states that management should document its internal controls—or processes management uses to help an entity achieve its objectives—and communicate quality information to all levels of the entity.[27] NASA's program management and systems engineering policy and implementing guidance do not require programs to develop such a plan, but most programs are only responsible for the mass of a single project or system. The Gateway program is integrating

[27] GAO-14-704G.

across multiple projects and systems. Therefore, it is important that the Gateway program communicates its overall mass management plan with the PPE, HALO, and DSL projects.

The Gateway program and its projects have many factors to consider when making decisions about how to reduce the CMV mass.

Documenting the Gateway program's overall mass management plan and communicating it internally within the program would help ensure that all involved parties agree on how to overcome the CMV mass challenge.

This would include ensuring that at the September 2024 critical design-informed synchronization review (1) the HALO project's specific mass reduction plan aligns with the program's overall mass management plan, and (2) all parties are aware of priorities for making key trade off decisions, including what the logistics vehicle will carry for the Artemis IV mission. Without documenting the mass management plan, the program risks the projects making individual decisions to reduce mass without considering how those decisions affect other projects and the overall mission design.

M2M-level risks and pending decisions. Several M2M-level risks and pending decisions could affect the Gateway program. For example:

- The M2M and Gateway programs are tracking risks related to mission operations that affect multiple programs. For example, the Gateway program will need to carefully manage where the visiting vehicles—such as a logistics vehicle, Orion, or a lunar lander—are docked and pointed throughout the mission. The Gateway and the visiting vehicles all have thermal, power, and communication needs, which require them to point in a certain direction at different times depending on lighting from the sun. If NASA cannot identify solutions to meet all visiting vehicle needs, it risks degraded performance for some vehicles, such as not generating enough power from solar arrays, or in a worst-case scenario, loss of mission. To mitigate this risk, visiting vehicles might have to undock with the Gateway and later redock over the course of the mission. For example, Gateway program officials said that after the crew integrates the International Habitat with the rest of the Gateway, they might have to move Orion and redock it in another position to ensure that it is facing in a direction to meet its thermal needs. The program office is conducting additional analysis in spring 2024 to determine the various vehicle needs.

- In addition, the M2M program has not yet determined how many revolutions in orbit the crew will spend on the Gateway. The higher the number of revolutions, the more consumables required on the Gateway for crew, which affects mass. As noted above, the Gateway program will have to make decisions about what it brings up on its logistics vehicle. The more consumables needed, the less mass that is available for other items.
- M2M program officials said they are continually assessing Artemis mission profiles and concepts of operations for crew safety.[28] As part of this assessment, M2M program officials said they consider scenarios on how they might change a mission profile if one of the programs is not ready on time. If NASA decides to make changes to its mission profiles, it could result in changes to planned time frames or the anticipated concept of operations for the Gateway program. As of April 2024, M2M program officials said they did not plan to make any changes.

It is important that the Gateway program ensures that the program and the PPE and HALO projects complete their efforts to mature technologies, improve design stability, and mitigate significant technical challenges and risks as the Gateway program moves toward its critical design-informed synchronization review in September 2024. This will help the program assess whether the overall Gateway design performs as expected. As part of the updated JCL at this review, the program will also be able to account for how these efforts affect the program's risk posture. For example, if the program documents its mass management plan and identifies ways to reduce the CMV mass without making design changes, it may reduce the associated consequences on cost, schedule, and performance before the PPE and HALO projects enter the system integration and test phase. Conversely, if program officials determine that they must make design changes to reduce mass, they can account for any associated cost growth or schedule delays in the JCL analysis.

---

[28] A mission profile is a brief description of the mission and its objectives. The concepts of operations are the more detailed plans for how the agency plans to use systems to achieve a mission's objective. For example, see the Artemis IV mission concept of operations depicted in figure 2.

NASA Plans Annual Reviews to Determine Roles for the Gateway in Future Mars Missions

NASA has held two Architecture Concept Reviews to map its high-level M2M exploration goals to the elements that are needed to achieve them. These two reviews largely focused on establishing the architecture review process and aligning existing systems, such as the Gateway, to the architecture. NASA plans to hold future reviews on an annual basis. Thus far, NASA has assigned several functions to the Gateway program that align to goals of the initial Artemis missions, like returning humans to the moon. NASA plans to use future reviews to determine if it needs to add capabilities to the Gateway to support the later segments, which will build on these early lunar explorations and eventually send a crew to Mars.

NASA's Initial Architecture Concept Reviews Focused on Early Lunar Missions

The Strategy and Architecture Office (SAO) plans to hold reviews in the November time frame each year to map NASA's high-level M2M exploration goals to the elements that are needed to achieve them. NASA held its first Architecture Concept Review in January and February 2023 and its second in November 2023. According to officials, the November time frame allows the Architecture Concept Review to inform the budget request for the following year. To establish this desired yearly cadence, NASA shortened the length of the first two review cycles. NASA plans for future cycles to last a full 12 months. In upcoming cycles, NASA plans to make numerous key decisions related to Mars missions in the areas of surface systems; entry, descent, landing, and ascent systems; transportation; and crew support.

NASA organized the M2M architecture into four segments that increase in complexity and mission scope over time:

- *Human Lunar Return:* the initial capabilities, systems, and operations necessary to re-establish human presence on and around the moon.
- *Foundational Exploration:* the expansion of lunar capabilities, systems, and operations supporting complex orbital and surface missions to conduct science and Mars precursor missions.

- *Sustained Lunar Evolution:* the enabling capabilities, systems, and operations to support science, economic opportunity, and a steady human presence on and around the moon.
- *Humans to Mars:* the initial capabilities, systems, and operations necessary to establish human presence and enable science and continued exploration on Mars.

The segmented approach allows NASA to incrementally develop and deploy elements as needed. It also allows NASA to break the architecture down into manageable pieces to prioritize its analysis work and coordinate with commercial, academic, and international partners.

According to SAO officials, the agency is building out the segments concurrently, as mission and segment operations may overlap. Therefore, it is not necessary to complete one segment before operations in the next segment begin. Officials said this approach allows them to prioritize areas that have a time sensitivity associated with them across all the segments.

Due to the abbreviated nature of the first two review cycles, those reviews largely focused on establishing the process and aligning existing systems to the architecture. In particular, the first review cycle and its associated products focused on the first segment of the architecture, Human Lunar Return. That review cycle included coordination with each of NASA's mission directorates. The second review cycle expanded on the first. It focused on refining and adding detail to the Human Lunar Return and Foundational Exploration segments, added elements that are further along in development, and developed strategies for making decisions about the eventual first crewed missions to Mars.

Within the four segments, NASA's architecture reviews have made the most progress in the Human Lunar Return segment in terms of mapping high-level objectives to more specific functions and allocating those functions to elements. Elements are flight programs, projects, and systems, such as the Gateway. NASA has also made progress in the Foundational Exploration segment, but to a lesser extent. NASA has not completed mapping objectives to specific elements for the Sustained Lunar Evolution. However, it completed some examples of notional mapping, such as for future habitation and transportation systems. NASA has not completed mapping objectives to elements for the Humans to Mars segment. See figure 7 for the four segments and the status of developing these segments as of January 2024.

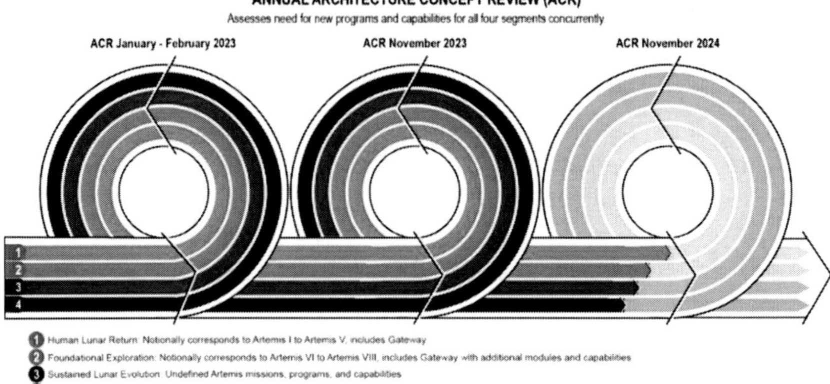

Source: GAO analysis of NASA documentation; GAO (illustration). | GAO-24-106878.

Note: According to Strategy and Architecture Office officials, the specific flight manifests, sequences, and specific mission content and design are notional and are subject to change due to factors such as budget, schedule, and other pressures that are beyond the scope of the architecture team.

Figure 7. Moon to Mars Architecture Segments as of January 2024.

As of the second architecture concept review in November 2023, NASA determined roles for the Gateway, as well as other existing Artemis systems, for the first two architecture segments—Human Lunar Return and Foundational Exploration. However, the agency had not yet determined roles for existing Artemis systems for the two later architecture segments—Sustained Lunar Evolution or Humans to Mars. For example:

- In the Human Lunar Return architecture segment, NASA determined that the Gateway will demonstrate the ability to conduct crewed lunar surface missions from cislunar space.[29] It will also provide capability for the physical assembly of spacecraft elements and crew habitation in cislunar space, among other contributions.
- In the Foundational Exploration architecture segment, the Gateway, including its international partner contributions, will provide a pressurized habitable environment in cislunar space for durations of months to years. The Gateway will also allow NASA to remotely operate the habitation system between crewed missions on the lunar

[29] Cislunar space is the volume of space around the moon featuring multiple possible stable staging orbits for future deep space missions.

surface and provide the capability to restore and stabilize the habitable environment.

Future Reviews Will Determine How Gateway Will Support Later Missions, Including to Mars

While NASA has not formally determined how the Gateway program will support the Sustained Lunar Exploration or Humans to Mars segments, the agency considers the Gateway a critical element of its M2M architecture. The SAO office is analyzing several potential uses of the Gateway as a proving ground for lunar surface activities and Mars development. For example:

- Operating an unattended system in deep space: The Gateway will spend most of its life uncrewed, for durations up to 3 years at a time, which is likely analogous for future Mars class systems.
- Long-term demonstration of a Mars transit habitat: The Gateway will provide the ability to stage long-duration microgravity systems in deep space or near-deep space equivalent environmental conditions and enable Mars-like analog missions closer to Earth.
- Crew transitions between micro-gravity and partial gravity: The Gateway's location in cislunar space will further NASA's understanding of crew transitions between microgravity and partial gravity environments to help NASA prepare for Mars exploration.
- Orbit-to-surface split crew operations: The Gateway will enable NASA to analyze conducting missions and station operations with some crew remaining on the Gateway while others descend to the lunar surface.
- Observations, science, and technology demonstrations in lunar orbit: The Gateway will enable progress toward Earth independence through applied science and investigations in areas such as atmospheric weather, space weather, and dust.

The SAO laid out several key decisions that it needs to make in upcoming review cycles and has started studies to inform those decisions. For example, the SAO plans to determine how it will transport crew to Mars, which could involve the Gateway. NASA plans to design a transit system to transport the crew; surface systems; and entry, descent, and landingsystems to deep space locations, including Mars, and to return crew back to Earth. NASA is considering the Gateway as an aggregation point—a location where NASA could physically assemble vehicles for missions in deep space—including to Mars. NASA officials told us that aggregating the Mars transit vehicle at the Gateway is likely the preferred option because its location in near rectilinear halo orbit is an advantage.[30]

SAO officials told us it would likely take several annual review cycles to fully determine the Gateway's role in later Mars missions. Thus, it could be several years before NASA fully understands the Gateway's role and contributions to these missions. Even if NASA identifies new capabilities for the Gateway, SAO and Gateway program officials said that they expect that they would be minor—such as software updates or moving the Gateway between orbits—and would likely not affect the initial capability. The January 2024 revised Architecture Definition Document notes that NASA should limit modifications to programs already in development. In the case of the Gateway initial capability, the PPE and HALO projects were in development for several years before the establishment of the Moon to Mars strategy. Design changes beyond this point would be costly to implement and could result in reduced capabilities. In addition, and as noted above, the projects are addressing significant mass concerns and other challenges, which could limit NASA's ability to add new capabilities to the Gateway's initial capability.

While significant design changes to the Gateway's initial capability are not feasible, NASA may be able to add capability to the Gateway over time.

NASA designed the Gateway for compatibility with the International Deep Space Interoperability Standards, which enable industry and international entities to independently develop compatible systems for deep space exploration.[31] Officials said the Gateway also provides a platform and docking ports for scalability to accommodate additions, such as a new module. As NASA continues to build out its architecture for later lunar and Mars missions, SAO and Gateway officials noted that the program would provide

[30] According to NASA, near rectilinear halo orbit is the optimal orbit to support the goals of the Artemis campaign because it provides low-cost, long-term stability, an environment achievable for vehicle designs, and accessibility for transportation elements. It also provides a favorable vantage point for Earth, sun, and deep space observations in a magnetically shielded environment, depending on the phase of the moon's orbit.

[31] NASA and the International Space Station partner agencies collaborated to develop the *International Deep Space Interoperability Standards*. The standards include nine discipline areas: avionics, communications, docking, environmental control and life support systems, power, rendezvous, robotics, thermal, and software. The Gateway is the first program to implement the standards.

input throughout the architecture concept review process on the Gateway's capabilities and constraints.

In our discussions with Gateway program officials, they noted two constraints that would affect the Gateway's ability to support Mars missions:

Stack controllability. According to Gateway program officials, stack controllability is the biggest constraint that the SAO and the Gateway program have discussed for future missions. As previously noted, if NASA decides to allow large, heavier visiting vehicles to dock with the Gateway, it affects the PPE's ability to control the stack. The program will need to reconsider the Gateway's mass capability calculations, including the size of the additional vehicle and its center of gravity, as NASA makes these decisions. In the case of a future Mars transit vehicle, officials said there are no designs currently mature enough for the Gateway program to determine how it will handle stack controllability. Officials said they would need to make adaptations on a case-by-case basis.

Life of the Gateway. The Gateway's planned on-orbit life of 15 years could also limit its use, depending on the timing of crewed Mars missions. Gateway program officials said they expect the Gateway to exceed its mission life as other systems have done. For example, NASA also designed the International Space Station for a 15-year mission life, which it has now exceeded by over 11 years. NASA expects to continue International Space Station operations until 2030. The Gateway, however, will operate in a different environment. It will be in cislunar space, where it will be exposed to higher levels of radiation than the International Space Station, which is in low Earth orbit. In addition, the Gateway will spend significant periods of time uncrewed between Artemis missions, while the International Space Station has been continuously crewed. The Gateway is planned to rely on numerous autonomous systems, such as robotics systems, to operate and maintain it between crewed missions and to prepare for the arrival of the next crew. However, to the extent that these systems fail or that the Gateway needs more maintenance than the autonomous systems can provide, the Gateway could be adversely affected. These factors could limit an extended on-orbit life. Further, independent studies requested by NASA show that the agency has explored launching Mars missions in 2039, though the studies noted several potential challenges to achieving this date. The Gateway program is currently working to launch its initial capability in 2027; thus, the Gateway could have exceeded its planned 15-year on-orbit life as early as 2042 when crewed missions to Mars are potentially just beginning.

NASA's ultimate goal with its lunar efforts is to prepare for crewed missions to Mars. The agency views the Gateway as a key asset in its efforts to create a sustainable lunar presence and test out technologies and concepts of operations for Mars. While the agency is past the point that it can make significant design changes to the Gateway, it may be able to add capability via software updates or by adding new modules. NASA plans to use its annual review cycles to identify ways it can more fully leverage the potential contributions of the Gateway, while balancing the Gateway's limitations, to inform its eventual plans for crewed Mars missions.

Conclusion

The Gateway program plans to launch the PPE and HALO together in 2027, in about 3 years. Between now and then, the projects will go through the riskiest phase of development—integration and test. Before they do so, the Gateway program will have an opportunity to assess its mission design and schedule at its planned September 2024 critical design-informed synchronization review. Addressing known risks and understanding their effect on the program is an important part of this process. The PPE and HALO projects have several significant challenges to overcome and risks to mitigate as they enter the integration and test phase, including reducing their mass to within their allocations. The program and its projects have plans to address some of these challenges prior to the review, such as conducting analyses related to stack controllability. However, the PPE and HALO projects are at the point where they are building hardware, and the program has not yet documented how to address their mass overages. As a result, NASA is at risk of building hardware that could need significant redesigns that would increase costs and delay schedule. Documenting a mass management plan and ensuring that all relevant parties can make timely decisions in alignment with that plan would be a key step for the program to ensure it has an executable mission design.

Recommendation for Executive Action

The NASA Administrator, in coordination with the Exploration Systems Development Mission Directorate, should ensure that the Gateway program documents its overall mass management plan and shares it with its projects ahead of the program's planned September 2024 critical design-informed synchronization review. (Recommendation 1)

Agency Comments and Our Evaluation

We provided a copy of this report to NASA for review and comment. NASA provided written comments that are reprinted in appendix III. In its response, NASA concurred with our recommendation and estimated it would take action to implement this recommendation in September 2024. NASA also provided technical comments that we incorporated as appropriate.

We are sending copies of this report to the NASA Administrator and interested congressional committees. In addition, the report is available at no charge on the GAO website at https://www.gao.gov.

If you or your staff have any questions about this report, please contact me at (202) 512-4841 or russellw@gao.gov. Contact points for our Offices of Congressional Relations and Public Affairs may be found on the last page of this report. GAO staff who made key contributions to this report are listed in appendix IV.

W. William Russell IV

William Russell
Director, Contracting and National Security Acquisitions

List of Committees

The Honorable Jeanne Shaheen
Chair

The Honorable Jerry Moran
Ranking Member
Subcommittee on Commerce, Justice, Science, and Related Agencies
Committee on Appropriations
United States Senate

The Honorable Frank Lucas
Chairman

The Honorable Zoe Lofgren
Ranking Member
Committee on Science, Space, and Technology
House of Representatives

The Honorable Hal Rogers
Chairman

The Honorable Matt Cartwright
Ranking Member
Subcommittee on Commerce, Justice, Science, and Related Agencies
Committee on Appropriations
House of Representatives

Appendix I: Objectives, Scope, and Methodology

This report examined (1) the Gateway program's plans for updating its cost and schedule analysis for its initial capability, (2) the extent to which the Gateway program has made progress with the Power and Propulsion Element (PPE), Habitation and Logistics Outpost (HALO), and the first Deep Space Logistics (DSL) vehicle needed to support the Artemis IV mission in 2028 and is addressing project risks, and (3) the National Aeronautics and Space Administration's (NASA) processes for determining how it will use the Gateway to support missions beyond Artemis IV, including Mars missions. This is the latest in a series of GAO reports addressing NASA's Artemis enterprise.[32] The focus of this report is on the Gateway program and its U.S.-led development projects.

[32] GAO, NASA Artemis Programs: Crewed Moon Landing Faces Multiple Challenges, GAO-24-106256 (Washington, D.C.: Nov. 30, 2023); NASA Lunar Programs: Improved Mission Guidance Needed as Artemis Complexity Grows, GAO-22-105323 (Washington, D.C.: Sept. 8, 2022); NASA Lunar Programs: Significant Work Remains, Underscoring Challenges to Achieving Moon Landing in 2024, GAO-21-330 (Washington, D.C.: May 26, 2021); NASA Human Space Exploration: Significant Investments in Future Capabilities Require Strengthened Management Oversight, GAO-21-105 (Washington, D.C.: Dec. 15, 2020); and NASA Lunar Programs: Opportunities Exist to Strengthen Analyses and Plans for Moon Landing, GAO-20-68 (Washington, D.C.: Dec. 19, 2019).

To understand the Gateway program plans to update the cost and schedule analysis for the initial capability, we reviewed the Gateway program's documentation, including key decision point (KDP) I review materials, and NASA's program and project management policy.[33] We also interviewed program officials to understand their methodology for and plans to update their joint cost and schedule confidence level (JCL) analysis and hold a KDP II review. To assess how the program's schedule, contracts, and risk posture have changed since the program conducted the JCL and might change when the program updates its JCL, we reviewed program and project documentation of top risks, program quarterly reviews, and KDP I review documentation. This included the results of the JCL and findings and recommendations from the program's independent standing review board. We also interviewed program and project officials on the KDP I review results, new risks and mitigation plans, schedule updates, and upcoming contract modifications.

In addition, to determine the extent to which the initial capability baseline and accelerated launch readiness dates support the Artemis IV mission, we reviewed the December 2023 KDP I decision memorandum that included the program's cost and schedule baselines for the initialcapability, contract documentation, quarterly program reviews, and risk information. To determine the date that the Gateway program would need to launch the PPE and HALO together to support the Artemis IV mission, we assessed program documentation and interviewed HALO project and Gateway program officials on the time needed for the initial capability to transit to orbit and check the orbit's stability prior to the start of the Artemis IV mission. We refer to this date as the need launch readiness date in the report. To determine the program's ability to support the Artemis IV mission date, we compared the need launch readiness date to the baseline launch readiness date. We interviewed Gateway program and HALO and PPE project officials to discuss their plans for determining an accelerated launch readiness date. We interviewed Moon to Mars (M2M) program officials to discuss their reassessment of Artemis programs and mission dates and their program and risk management approach for the Artemis programs.

To determine the extent to which the Gateway program has made progress with the PPE, the HALO, and the first DSL vehicle needed to support the Artemis IV mission in 2028, and is addressing project risks, we (1) analyzed program documents, (2) determined if the projects met our best practices for

[33] NASA, NASA Space Flight Program and Project Management Requirements, NASA Procedural Requirements 7120.5F (Aug. 3, 2021).

technology readiness and design stability, and (3) assessed the projects' progress against NASA's policies and guidance. More specifically, we analyzed Gateway program and project documentation that contained data on the projects' accomplishments, schedules, designs, technical challenges, and risks. This included documentation and data from the Gateway program's quarterly reviews from April 2023 to February 2024, KDP I review, and the June 2023 HALO and March 2024 PPE project critical design reviews. We also analyzed documentation from the Solar Electric Propulsion (SEP) project, including monthly review slides from May 2023 to April 2024, to determine the project's progress on maturing the technology for the PPE's flight thrusters and the status of building the hardware. We compared the progress of the Gateway program and its projects against the NASA acquisition life cycle in NASA's policy for program management for space flight projects.[34]

To determine if the projects met our best practices for technology readiness and design stability, we obtained data from the PPE and HALO projects on the technology readiness levels of their critical technologies at preliminary design review and the percentage of design drawings released at critical design review through a data collection instrument. We then discussed this in interviews with project officials and compared these data against our best practices.[35] We interviewed Gateway program and PPE, HALO, DSL, and SEP project officials on the status of designing and developing hardware and software and to discuss the results of their technical reviews.

To assess the projects' progress toward mitigating technical and design risks and challenges, such as the projects exceeding their mass allocations, we compared program and project documentation of top risks and challenges and mitigation plans against NASA systems engineering policy and guidance and NASA's leading indicators reference guide.[36] We interviewed Gateway program and PPE, HALO, DSL, and SEP project officials to discuss the status of developing hardware and software, to program and project risks, and risk mitigation plans. In addition, we assessed M2M risk information and status

[34] NASA, *NASA Space Flight Program and Project Management Requirements*, NASA Procedural Requirements 7120.5F (Aug. 3, 2021).

[35] GAO, Best Practices: Using a Knowledge-Based Approach to Improve Weapon Acquisition, GAO-04-386SP (Washington, D.C.: Jan. 1, 2004); and Best Practices: Capturing Design and Manufacturing Knowledge Early Improves Acquisition Outcomes, GAO-02-701 (Washington, D.C.: July 15, 2002).

[36] NASA, NASA Systems Engineering Handbook, NASA/SP-2016-6105 Rev 2 (2016) NASA Common Leading Indicators Detailed Reference Guide (January 2021); and NASA Systems Engineering Processes and Requirements, Procedural Requirements 7123.1C (Feb. 14, 2020).

review slides to identify any M2M level risks or decisions that might affect the Gateway program's progress. We interviewed M2M program officials to discuss those risks and decisions, as well as the Artemis IV concept of operations.

We also determined that the control environment and information and communication components of federal standards for internal control were applicable to our second objective.[37] To evaluate the control environment, we determined that the documentation of the internal control system principle was applicable. To evaluate NASA's control environment, we assessed Gateway program documentation and interviewed program officials on their plans to document key controls, including how they planned to manage mass for the comanifested vehicle. For the information and communication component, we determined the principle that management should internally communicate quality information to achieve an entity's objectives was applicable. To evaluate thiscomponent, we assessed Gateway program documentation, including documentation of program and project updates and risk management plans, and interviewed program and project officials to determine how the program communicated risk information internally, including with its projects.

To examine NASA's processes for determining how it will use the Gateway to support missions beyond Artemis IV, including Mars missions, we reviewed documentation of NASA's architecture concept review process—the process the agency is using to map high-level M2M objectives to the specific elements that will support science and exploration goals. More specifically, we reviewed documentation from NASA's early 2023 and November 2023 architecture concept reviews and strategic analysis cycles including the Architecture Definition Document, white papers, architecture summary documents, and briefing slides. We also reviewed M2M and Gateway program documents including risk management and requirements documents.

To understand how NASA defines systems architecture and how NASA has organized itself to develop and oversee its M2M architecture, we reviewed NASA systems engineering policies and guidance; program and project management policies and guidance; and documentation of the Exploration Systems Development Mission Directorate's organizational structure. We interviewed M2M program officials to discuss updates to the agency

[37] 6GAO, Standards for Internal Control in the Federal Government, GAO-14-704G (Washington, D.C.: September 2014).

organizational structure and how the M2M program coordinates with other offices. We interviewed Strategy and Architecture Office (SAO) officials to discuss the strategic analysis cycle and architecture concept review process, the extent to which the Gateway's role in missions beyond Artemis IV has been determined, how the office coordinates with internal and external stakeholders, and the office's plans and expectations to fully define the Gateway's role in such missions. We also interviewed Gateway program officials to discuss their participation in the architecture concept review process, ongoing trade studies, and the extent to which new requirements had been levied on the program as a result of these new processes, if at all. We also discussed NASA's potential plans to use the Gateway as an assembly point for a future Mars transportation vehicle and limitations of the Gateway with both SAO and Gateway program officials.

We conducted this performance audit from May 2023 to July 2024 in accordance with generally accepted government auditing standards. Those standards require that we plan and perform the audit to obtain sufficient, appropriate evidence to provide a reasonable basis for our findings and conclusions based on our audit objectives. We believe that the evidence obtained provides a reasonable basis for our findings and conclusions based on our audit objectives.

Appendix II: NASA's Annual Architecture Concept Review Process

NASA created the Architecture Concept Review process to map high- level objectives to the specific elements that will support science and exploration goals. The process centers on an annual study cycle—called the Strategic Analysis Cycle. During this analysis cycle, NASA continually updates and refines the architecture, incorporating feedback from stakeholders from within NASA and across industry, academia, and international partners. Each analysis cycle culminates in an annual Architecture Concept Review—a review that brings together NASA leadership to refine the existing architecture and strategies.

NASA's architecture concept review process is based on "architecting from the right, executing from the left."[38] The agency begins by tracing the broadest, most long-term goals that are furthest in the future on the timeline from NASA's high-level objectives and then works backward from that goal to establish the complete set of required elements. Currently, the driver for the overall architecture is planning for a human mission to Mars. Meanwhile, systems and elements are "executed from the left" in a regular development process, integrating systems as they move left to right within the architecture (see fig. 8).

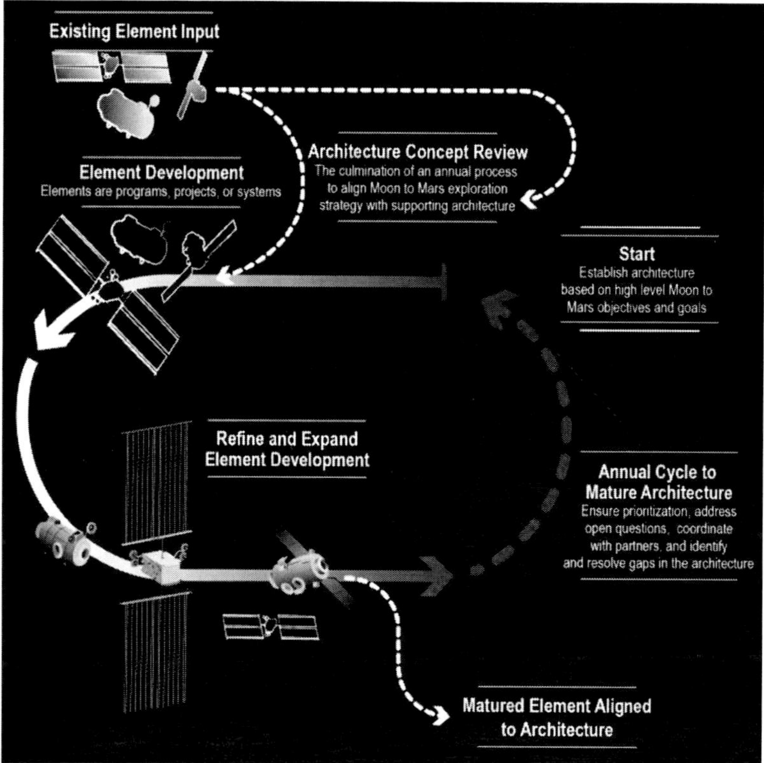

Source: GAO analysis of NASA documentation; GAO (illustration). | GAO-24-106878.

Figure 8. Notional Architecture Concept Review Process.

---

[38] NASA, Exploration Systems Development Mission Directorate, *2023 Moon to Mars Architecture Definition Document (ESDMD-001) Revision A,* NASA/TP-20230017458 (Washington, D.C.: January 2024).

NASA's Architecture Definition Document states that the agency is using an applied systems engineering method to facilitate applying these principles to the architecture definition. The first part of this method is an ordered process of objectives' decomposition to complete the process of architecting from the right. In this process, NASA officials identify the characteristics and needs to assure objective satisfaction. They then trace these characteristics and needs to the functions and use cases that elements and systems must accomplish.[39] The second supporting method is establishing an architectural framework to organize, integrate, and track the allocation of functions and use cases to the executing programs. This structure is intended to enable the integration of the system-of-systems development, identify gaps in the architecture, and adjust the architecture as left-to-right execution occurs, technologies mature, or objectives are satisfied.

New programs may formally enter the architecture via a new milestone the Strategy and Architecture Office has implemented, called Element Initiation. During Element Initiation, the mission directorate reviews whether a proposed element provides a solution to needs or gaps identified in the Moon to Mars architecture and determines whether to apply necessary resources to formulate that element. Element Initiation occurs in pre-formulation and moves programs that the Architecture Concept Review process initiated into the traditional NASA space flight program development life cycle. NASA initiated two new elements through this process in 2023: (1) a lunar surface habitation element, which may provide an initial home for astronauts on the moon, and (2) a lunar surface cargo lander element, which may deliver supplies and equipment to the moon.

[39] NASA's Architecture Definition Document defines use cases as operations that would be executed to produce the desired needs and/or characteristics. It defines functions as actions that an architecture would perform to complete the desired use case.

Appendix III: Comments from the National Aeronautics and Space Administration

National Aeronautics and Space Administration

Mary W. Jackson NASA Headquarters
Washington, DC 20546-0001

Reply to Attn of: Exploration Systems Development Mission Directorate

Mr. W. William Russell
Director
Contracting and National Security Acquisitions
United States Government Accountability Office
Washington, DC 20548

Dear Mr. Russell:

The National Aeronautics and Space Administration (NASA) appreciates the opportunity to review and comment on the Government Accountability Office (GAO) draft report entitled, "NASA Artemis Programs: NASA Should Document and Communicate Plans to Address Gateway's Mass Risk" (GAO-24-106878), dated June 13, 2024.

NASA plans to build a sustained human lunar presence and ultimately travel to Mars through a series of flights known as Artemis. The Agency is developing the Gateway—a space station planned to orbit the Moon in support of lunar exploration. NASA committed to launching the Gateway initial capability by December 2027 in support of the Artemis IV flight. The launch will include the first components of the Gateway—the Power and Propulsion Element and the Habitation and Logistics Outpost.

NASA's Exploration Systems Development Mission Directorate (ESDMD) acknowledges the crucial role played by GAO in assessing instances of waste, fraud, or abuse within the Federal Government. ESDMD is dedicated to transparency and accountability. The Agency's goal is to fully cooperate with GAO, providing access to all relevant information and documentation necessary for its audits, evaluations, and investigations. During this audit, ESDMD provided 85 products and attended 14 requested interviews.

NASA appreciates GAO's understanding that development projects inherently operate in dynamic environments, marked by frequent changes, updates, and iterations. The landscape can swiftly evolve during the development cycle, posing difficulties in aligning with evolving requirements, timelines, and deliverables. Additionally, parallel development and resultant changing requirements are an inherent reality that requires adaptability and flexibility from all involved parties. As projects progress, stakeholders gain deeper insights necessitating adjustments to initial specifications. Effective management of changing requirements involves clear communication, diligent documentation, and collaborative decision-making to assess the impact on timelines, resources, and project scope. While accommodating changes may introduce challenges such as increased development time or additional costs, addressing evolving needs proactively can lead to a more robust final product.

In the draft report, GAO states that the current Gateway schedule baseline does not align with the September 2028 Artemis IV flight date; however, the program has a plan to support the Artemis IV flight. ESDMD is committed to maintaining accountability among contractors regarding project timelines, even in the face of schedule slippage of other elements. Adhering to agreed-upon dates for all contracts fosters discipline and ensures that all stakeholders are committed to meeting project milestones. Holding contractors accountable reinforces the importance of timely delivery and helps mitigate further delays by instilling a sense of urgency. Additionally, it enables project managers to identify bottlenecks early on and take corrective actions to realign the project trajectory. While flexibility may be necessary in certain circumstances, maintaining firm expectations for adherence to dates encourages a culture of responsibility and ultimately contributes to the overall success of the project. To effectively manage the Gateway program, NASA is committed to proactively addressing potential risks and the Agency Baseline Commitment. By prioritizing an earlier delivery date, NASA aims to ensure the program's prompt completion. This approach underscores our determination to mitigate identified risks and uphold our commitment to delivering the Gateway as efficiently as possible.

NASA uses an array of tools to track quality, progress, and performance against cost and schedule targets, which include, but are not limited to, Government mandatory inspection points, project-level cost and schedule joint confidence level informed commitments (including for major developmental upgrades), independent reviews at major life-cycle reviews and associated key decision points, documented and configuration-controlled mission definition baselines, risk assessments, independent Agency financial auditing (including a thirteenth consecutive unmodified or "clean" audit opinion in 2023), and Agency-led baseline performance and major program reviews. There are also independent reviews by the NASA Advisory Council, Aerospace Safety Advisory Panel, and various other ongoing reviews from Governmental oversight entities. This rigorous monitoring helps NASA maintain accountability and quality in its programs and projects.

The Artemis implementation is unique from other NASA activities in that the flexible architecture is a guiding principle within the Artemis Campaign, enabling NASA to adapt to changing requirements, leverage partnerships, and achieve sustainable and cost-effective human exploration of the Moon and beyond. By embracing flexibility and innovation, NASA aims to establish a robust infrastructure and lay the foundation for future exploration missions to Mars and beyond. The approach NASA is pursuing ensures that capabilities are developed to meet the needs of the architecture.

In the draft report, GAO makes one recommendation addressed to the NASA Administrator.

Specifically, GAO recommends the following:

Recommendation 1: The NASA Administrator, in coordination with ESDMD, should ensure that the Gateway program documents its overall mass management plan and shares it with its projects ahead of the program's planned September 2024 critical design-informed synchronization review.

Management's Response: NASA concurs with this recommendation.

3

The Gateway program is working closely with both its prime contractors for initial capability, Northrop Grumman and Maxar Space Systems (Maxar), to ensure the final mass of the Co-Manifested Vehicle aligns with the performance of the launch vehicle such that the spacecraft can perform a timely transit to near-rectilinear halo orbit with sufficient capability to meet mission requirements. Both Northrop Grumman and Maxar have developed internal processes to manage the mass of their respective spacecraft, and the Gateway program is actively working to document an encompassing mass management plan, placing special emphasis on cross system mass management and mission design.

The Gateway program is currently approaching the next major milestone in its life-cycle, the Critical Design Review (CDR), where it will evaluate the integrity of the project design and its ability to meet mission requirements with appropriate margins and acceptable risk. The Gateway program plans to finalize its mass management plan prior to CDR. To help ensure this plan is comprehensive, a preliminary version is already being used to supplement the internal mass management processes of both contractors. Their feedback, along with any lessons learned, will be incorporated into the final revision to ensure an objective decision-making framework exists for mass management throughout the remainder of Gateway's development.

Estimated Completion Date: September 30, 2024.

We have reviewed the draft report for information that should not be publicly released. As a result of this review, we have not identified any information that should not be publicly released.

Once again, thank you for the opportunity to review and comment on the subject draft report. If you have any questions or require additional information regarding this response, please contact Christine Solga at (202) 358-1238.

Sincerely,

NED PENLEY Digitally signed by NED PENLEY Date: 2024.07.15 09:36:30 -05'00'

Catherine A. Koerner

Appendix IV: GAO Contact and Staff Acknowledgments

GAO Contact

William Russell, (202) 512-4841 or RussellW@gao.gov

Staff Acknowledgments

In addition to the contact named above, Kristin Van Wychen (Assistant Director); Erin Kennedy (Analyst-in-Charge); Lena Burleson; Susan Ditto; Lorraine Ettaro; Laura Greifner; Tonya Humiston; Jason Lee; Joe Mancina; Carrie Rogers; Sylvia Schatz; Kate Sharkey; Kevin Walsh; Alyssa Weir; and Robin Wilson made significant contributions to this report.

GAO's Mission

The Government Accountability Office, the audit, evaluation, and investigative arm of Congress, exists to support Congress in meeting its constitutional responsibilities and to help improve the performance and accountability of the federal government for the American people. GAO examines the use of public funds; evaluates federal programs and policies; and provides analyses, recommendations, and other assistance to help Congress make informed oversight, policy, and funding decisions. GAO's commitment to good government is reflected in its core values of accountability, integrity, and reliability.

Connect with GAO

Connect with GAO on Facebook, Flickr, Twitter, and YouTube. Subscribe to our RSS Feeds or Email Updates. Listen to our Podcasts. Visit GAO on the web at https://www.gao.gov.

To Report Fraud, Waste, and Abuse in Federal Programs

Contact FraudNet:

Website: https://www.gao.gov/about/what-gao-does/fraudnet Automated answering system: (800) 424-5454 or (202) 512-7700.

Congressional Relations

A. Nicole Clowers, Managing Director, ClowersA@gao.gov, (202) 512-4400, U.S. Government Accountability Office, 441 G Street NW, Room 7125, Washington, DC 20548.

Public Affairs

Sarah Kaczmarek, Acting Managing Director, KaczmarekS@gao.gov, (202) 512-4800, U.S. Government Accountability Office, 441 G Street NW, Room 7149 Washington, DC 20548.

Strategic Planning and External Liaison

Stephen J. Sanford, Managing Director, spel@gao.gov, (202) 512-4707
U.S. Government Accountability Office, 441 G Street NW, Room 7814, Washington, DC 20548.

Chapter 5

Economic Growth and National Competitiveness Impacts of the Artemis Program*

Patrick Besha

Acknowledgements

Wish to thank Alex Macdonald and Phil McAlister for their insightful reviews of this report, and Elaine Gresham for copyediting and graphics support.

Economic Growth and National Competitiveness Impacts of Artemis

NASA's Artemis Program will return humans to the lunar surface in the 2020s, create a sustainable American presence for years to come, and provide the experience and technology to conduct the first human missions to Mars. While providing scientific returns, geopolitical leadership, and the enduring value of human and robotic space exploration, the program also offers the prospect of substantial economic growth and enhanced U.S. national competitiveness.

NASA funds commercial lunar initiatives that follow in the footsteps of the successful Commercial Orbital Transportation Services (COTS), Commercial Resupply Services (CRS), and Commercial Crew (CCrew)

---

* This is an edited, reformatted and augmented version of Office of Technology, Policy & Strategy (OTPS) Publication, dated May 2022.

In: Back to the Moon
Editor: Silas Blackwood
ISBN: 979-8-89530-861-5

programs that revolutionized space transportation, cargo supply to low-Earth orbit, and human spaceflight. The Commercial Lunar Payload Services (CLPS) program, Human Landing System (HLS) program, and NextSTEP Broad Agency Announcement (BAA) lunar initiatives seek to fund commercial and private companies to deliver payloads, services, and astronauts to the lunar surface [1]. Much like earlier commercial programs, the intention is to grow U.S. commercial technical capability, broaden the space sector, and provide the impetus for innovation.

This study summarizes the economic and national competitiveness benefits of the Artemis program to complement its inherent values of scientific discovery, technology development, space exploration and global leadership. It includes a brief background on the success of NASA's historical commercial space initiatives and how Artemis program is an extension of COTS and Commercial Crew. It also examines how the Artemis program may benefit the U.S. economy by driving innovation and new capabilities, lowering the cost of access to space, improving market competition and expanding the potential customer base. Finally, it provides an assessment of the Artemis program's role in helping U.S. firms compete in the global marketplace and the role of government in funding early-stage technology development. This study primarily focuses on the economic and national competitiveness impacts of the commercial elements of the program, and suggests they represent an evolution of the COTS/CCrew programs.

Additional perspective on the significant impacts of all Moon to Mars programs (including ISS, SLS, Orion, Gateway, HLS, CLPS, etc) is referenced in the appendix.

Key Highlights

- NASA's selection of the SpaceX Starship vehicle for a human lunar landing mission helps build a new commercial launch capability. Due to its size, this vehicle may lower launch costs to orbit significantly, expand the potential customer base for launch services, and increase U.S. market share.
- The CLPS program provides service contracts to companies building robotic lunar landers with a lower cost and shorter development schedule than previous missions. These companies also offer payload delivery services to commercial and international customers. CLPS

supports the development of a new, multibillion-dollar space subsector in lunar surface operations.

- Artemis drives innovation with challenging program requirements and its "mission-driven" approach. Potential solutions for lunar surface missions include technologies in compact power systems, synthetic biology, additive manufacturing, on-orbit fuel transfer, lunar surface mobility, and materials science.
- Economic activity from "Moon to Mars" programs across the U.S. includes:
    - More than 69,000 jobs
    - Spending that supports more than $14 billion in total economic output
    - An estimated $1.5 billion in federal, state, and local tax revenues nationwide
- National competitiveness impacts include the diffusion of new space technology more broadly across U.S. space enterprise, expansion of the U.S. technical workforce and STEM talent pipeline, enhancement of U.S. soft power, and stronger capacity of U.S. space firms to compete globally.

Commercial Space Programs in Perspective

The COTS program enabled U.S. private industry to develop space transportation capabilities with the goal of safe, reliable, and cost-effective access to low-Earth orbit. Ultimately, this program delivered two new low-cost U.S. launch vehicles carrying cargo to and from the ISS: SpaceX Falcon 9 and Orbital Antares. The vehicles can also be used at a cost-savings advantage for small and mid-sized NASA Science missions (e.g., Transiting Exoplanet Survey Satellite, Jason-3, Sentinel-6 Michael Freilich). Riding on the success of the Falcon 9, SpaceX grew its share of the global commercial launch market from 0% in 2011 to 78% in 2020 (and over 50% of overall launch market worldwide in 2021, excluding China).

SpaceX also jumpstarted the nanosatellite and CubeSat markets.

With the success of the launch vehicles, NASA entered into major multibillion-dollar commercial cargo resupply services (CRS) contracts with the two firms, creating future robustness for ISS cargo transportation. Building further upon these programs, NASA initiated the CCrew program for private sector firms to carry U.S. astronauts into orbit. To date, SpaceX has launched four crewed missions to the ISS (with several additional missions planned). A second competitor, Boeing's Starliner vehicle, is under development and slated for a crewed demonstration mission in 2022.

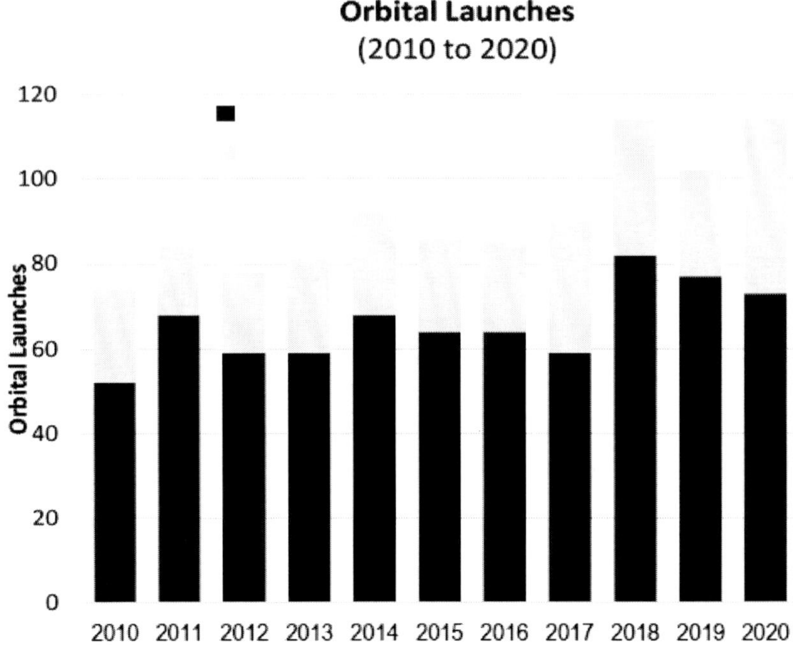

Source: BryceTech.

Commercial launches include internationally competed launches, any launch in which the primary spacecraft and launch vehicle are commercially owned or operated, and test launches of commercial launch vehicles. Launches procured under COTS, CRS, or CCrew are commercial.

Economic Impacts of the Artemis Program

It's important to note that the Artemis program rebuilds capabilities not present in the U.S. space enterprise since the Apollo era. In many cases, a capability has not existed for decades, the production knowledge has disappeared, and new technology must be invented.

Artemis offers the chance to reconstitute the American moon program, which includes super-heavy launch vehicles (SLS, Starship); Earth-return-capable crewed spacecraft (Orion); lunar orbit docking; lunar landing system (Starship HLS); and lunar surface EVA suits. In addition, sustainable human presence requires wholly new elements, including the Gateway, on-orbit fuel transfers, and lunar surface habitation and expanded mobility capabilities. One prominent commercial example central to Artemis is the SpaceX Starship HLS. Another is the Gateway Logistics Services.

NASA intends the initial Human Landing System contract ($2.9 billion) to procure payload delivery of multiple tons of cargo and astronauts to the surface of the moon in the 2020s. The selection of the in- development Starship rocket and HLS crew lander to deliver U.S. astronauts to the lunar surface can have far-reaching impacts. The following sections describe how Artemis can expand economic growth similarly to previous commercial space programs (COTS, CRS, CCrew).

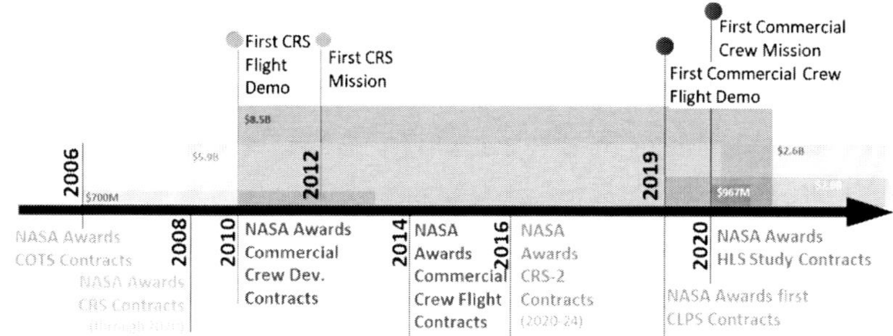

Source: NASA OIG Report data.

Timeline of major NASA Human Exploration Commercial Contracts.

New Capabilities and Lower Cost Access to Space

- Starship HLS is to provide a cargo delivery and astronaut landing capability for NASA. Other potential revenues in the future include lunar tourism and payload delivery (estimates are highly variable and range from several million to over a billion dollars) [2]. The ability to deliver tons of cargo, which may include scientific and commercial payloads, concurrently with astronaut landings offers a major rideshare opportunity to advance scientific, diplomatic, and economic development aims. In addition, broader strategic government interests in cislunar space may become possible with this new capability.

- Just as with the ISS, the regular provision of cargo and payload delivery underpins the future long-term establishment of a robust anchor tenet. The anchor for Artemis is infrastructure on the surface of the moon and in lunar orbit (Gateway). Due to the expense and complexity involved in these missions, NASA and international partners will be the first customers, but the customer base may eventually evolve towards stronger commercial and private sector participation. The initial steady stream of government contracts can ensure new capabilities and services are maintained until they reach commercial profitability.
- Starship's lift capability (> 100t to LEO) is significantly greater than existing commercial launch vehicles (SLS Blk 2 will have a similar capability). As larger vehicles have come on line, the cost/per kg to orbit has dropped for many applications [3]. Lowering the cost of access to space has historically opened up new space sectors, including nanosatellites, CubeSats, Earth Observation, and communications satellite constellations, as well as the development of larger, more sophisticated satellites.

Source: NASA OIG.

- Low-cost access to space has inverted the historical supply-demand curve in the space launch market. Where once high-cost suppliers had trouble fulfilling demand, today demand outstrips available low-cost launch supply. Consequently, entirely new customers and architectures in orbit become possible: from universities and private citizens to Earth observation CubeSats and LEO-based internet constellations. While commercial demand drivers for cislunar and lunar surface appear nascent now, as costs decrease and opportunities increase, the demand is likely to change. The figure (right) shows the proliferation of lunar companies worldwide across multiple subsectors, including entirely new commercial interests such as ISRU.

Market Competition and Improved Value to Government

- The initial procurements for CLPS, HLS, and other NextSTEP BAA awards were broadly competed, with selections based on technical, cost, and management factors [4]. There are fourteen eligible vendors for CLPS contracts, with major task order awards to four lunar payload service companies. Similarly, while a single firm received an HLS contract, multiple firms competed and are eligible to compete for follow-on missions. Market competition supports the broadening of expertise across the industry and generally drives per-mission costs down. Robotic lunar payload services and crewed astronaut landing support are now multibillion-dollar space subsectors.
- Firm competition showcases a strength of the American economy and system of government, drawing a notable contrast to the space programs of military-controlled and state capitalist systems.

Improving Value to the Government Through Innovative Contracting

By awarding a firm-fixed price contract, NASA departed from traditional fee-plus contracts that have seen program costs spiral. Elements of the cislunar Gateway platform have also followed this new contract model. If successful, the new model may lead to procurement reforms to improve government efficiency, enabling better distribution of resources to other elements of the Artemis program.

Program Characteristics	Traditional Approach	Commercial-Oriented Approach
Owner	NASA	Industry
Contract type	Cost-plus	Fixed price
Contract management	Prime contractor	Public-private partnership
Customers	NASA	Government and nongovernment
Funding for capability development	NASA procures capability	NASA provides "milestone" payments as agreed goals are reached
NASA's role in capability development	NASA defines "what" and "how"	Industry defines "how" and NASA approves
Requirements definition	NASA defines detailed requirements	NASA defines only higher-level requirements
Cost structure	NASA incurs total cost	NASA and industry share costs

Source: NASA Space Operations Mission Directorate / Exploration Systems Development Mission Directorate.

Expansion of Customer Base for Lunar Flights, Services, and Payload Delivery

- The U.S. Department of Defense (DOD) expressed interest in procuring payload services for upcoming commercial flights to and around the moon. Given the utility of this domain and the funding capacity of the customer, DOD may eventually be an important source of revenue.
- In the last few years, the roster of countries interested in fielding robotic missions to the moon grew to include UAE, Israel, S. Korea, and India. Countries interested in providing instruments for upcoming lunar missions include Germany, France, Saudi Arabia, and others. Astrobotic and other companies have acknowledged they maintain significant pending contracts for lunar payload delivery services from over a dozen countries (although for many customers there may be an uncertain ability to fulfill these commitments). A greater revenue base for American lunar companies provides financial robustness and market certainty, allowing them to grow their operations and create more high-paying U.S. jobs.

Commercial Human Spaceflight

- The CCrew program and successful demonstration of carrying astronauts to the ISS paved the way for the recent all-civilian Inspiration4 mission to orbit [5]. Using the Dragon crew capsule (reused from Crew-1 mission to the ISS) and the Falcon 9 rocket, the Inspiration4 mission was commercially procured by a private citizen and featured a charity fundraising element. After years of apparent stagnation, private astronaut missions are quite literally taking off. In addition to Inspiration4, Space Adventures flew a Japanese billionaire to the ISS in late 2021; one mission chartered by Axiom brought private citizens to the ISS in April 2022; Russia's Roscosmos lofted a filmmaker and leading actor to the station to film a movie production in October 2021; Blue Origin has flown three commercial missions; and Virgin Galactic has flown one mission to the edge of space. The estimated per-flight cost of these missions varies from several hundred thousand dollars to over $70 million, depending on the vehicle, destination, and duration.

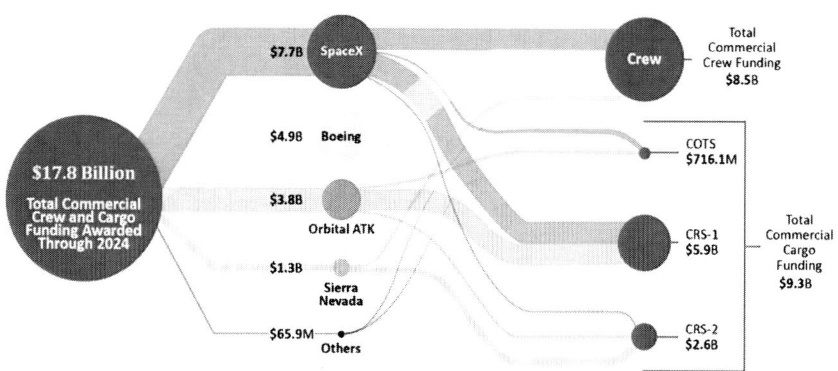

Source: NASA OIG analysis of Agency commercial activities and awards.

Driver of Innovation

- NASA's lunar programs are largely based on the model, processes, and contractual agreements pioneered by commercial space programs.[1] Such investments in private sector R&D have historically yielded significant advances in launch technology, avionics, and spacecraft design not foreseen or planned by the government (such as reusable rockets). The complexity of the human landing system and lunar surface operations is likely to generate advances in technology similar to how the Apollo missions advanced state-of-the-art in materials science, microelectronics, life support systems, and other areas. The figure (*right)* illustrates significant momentum of corporate and private sector investment in new space companies. This financing enables companies to develop innovative new hardware/software, generate revenue, become profitable and contribute to national economic growth. It may also help companies broaden their customer base beyond government to encourage greater stability and financial robustness.

- In a "moonshot" approach to innovation, high-technology R&D historically has greater economic impacts than many other areas of public sector investment [6].
- Technology spinoffs have historically been a significant benefit of government investment in basic and applied R&D, broadly diffused to U.S. companies.
    - Currently, the need for power on the surface of the moon is generating interest and investment in small, modular reactors. If established as a viable technology, these reactors would have tremendous variety of terrestrial applications and could provide for populations presently underserved. (See, for example, a startup developing this technology.) [7].
    - Food for space explorers may be developed using synthetic biology, given the absence of traditional organic ingredients and agricultural methods in space [8]. NASA's SynBio project aims to find a pathway to convert carbon dioxide into organic compounds and enable biomanufacturing in space. Such technologies could lead to the development of long shelf-life foods, new and more sustainable techniques for creating medicines and plastics, and more efficient and less wasteful global agricultural practices.

---

[1] NASA estimates that the development of Falcon 9 cost $400 million, or roughly ten times less than if built under traditional government contracting.

- o Additive manufacturing in space, including 3D-printed solar panels on the lunar surface, may yield manufacturing advances that can both accelerate the creation of space infrastructure and be utilized on Earth. The establishment of infrastructure is crucial to expanding human presence in Earth orbit, cislunar space, and the lunar surface. In-space manufacturing may enable platforms and satellites at much larger scales than what can be launched into orbit [9]. Already, in-space manufacturing on the ISS yields ultra-high- fidelity fiber optics (e.g., ZBLAN) and protein crystal growth in microgravity that surpasses the quality produced in terrestrial labs. Using regolith on the moon to build structures, including protective shields against solar radiation, may ultimately enable long-term presence on the moon. Such a foundational technology also has terrestrial applications, as remote populations and resource exploration platforms can benefit greatly from the ability to manufacture on-demand tools, parts, and other items.
- o The ISS is an important testbed and demonstration platform for Artemis missions. Over its roughly 20-year lifetime, it has yielded a number of significant breakthroughs with applications in medicine, physics, chemistry, materials science, and other areas [10].

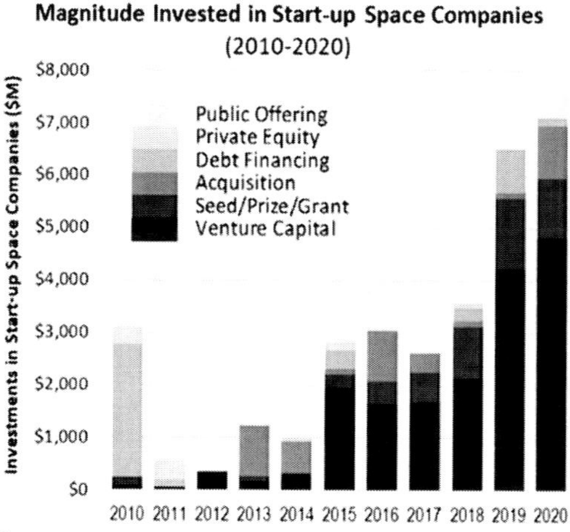

Source: BryceTech.

Space start-up ventures are any space firm that has received and reported seed funding or venture capital.

Broad Economic Impacts

- The broad Moon to Mars effort employs thousands of highly skilled professionals, and the U.S. Government channels billions of dollars into states through its contracts.
- M2M's economic impact goes beyond its immediate employment of NASA civil servants. The program creates wide-ranging benefits for local and state economies, as well as the U.S. economy, as NASA's contracting boosts economic activity elsewhere in the country: [11].
  - o Economic activity across the U.S. supports more than 69,000 jobs.
  - o Spending supports more than $14 billion in total economic output across the country.
  - o M2M programs generate an estimated $1.5 billion in federal, state, and local tax revenues nationwide.
  - o The supplier base for major programs, such as SLS, is spread out over all 50 U.S. states, supporting initiatives such as "Buy American" and strengthening our domestic supply chain.

National Competitiveness Impacts of the Artemis Program

National competitiveness is a measure of a nation's firms' ability to compete in the global marketplace. Competitiveness depends on the ability of firms to innovate and upgrade, and claim greater market share. Different factors contribute to this ability. For commercial lunar firms and the Artemis program, the most important factors are technology development, workforce, market conditions, and government role. U.S. economic growth has slowed in recent decades, largely due to globalization, as technology-based economic competitors take ever greater shares of high value-added markets at the expense of U.S. domestic industries [12]. To remain competitive, U.S. firms need to maintain a world-class technical workforce and sustain robust investment in technology.

CLPS companies such as Astrobotic and Intuitive Machines and HLS companies such as SpaceX develop technologies that do not currently exist commercially, including advances in low-cost manufacturing and spacecraft design. These technologies may ultimately diffuse to other space companies. NASA's recent missions to lunar orbit (LRO, LADEE) cost ~$250 to $583 million and did not feature a lander, generally a more costly and complex element. In contrast, cost estimates for the first CLPS lander missions are ~$80 million, and ~$200 million for the VIPER surface rover mission. The diffusion and rapid maturation of technology is at the core of the commercial lunar effort, as NASA seeks to harness private sector innovation to speed up mission development and lower costs. Furthermore, supporting entrepreneurship and incorporating new companies into high-tech supply chains broadens areas for innovation and can make supply chains more resilient and robust.

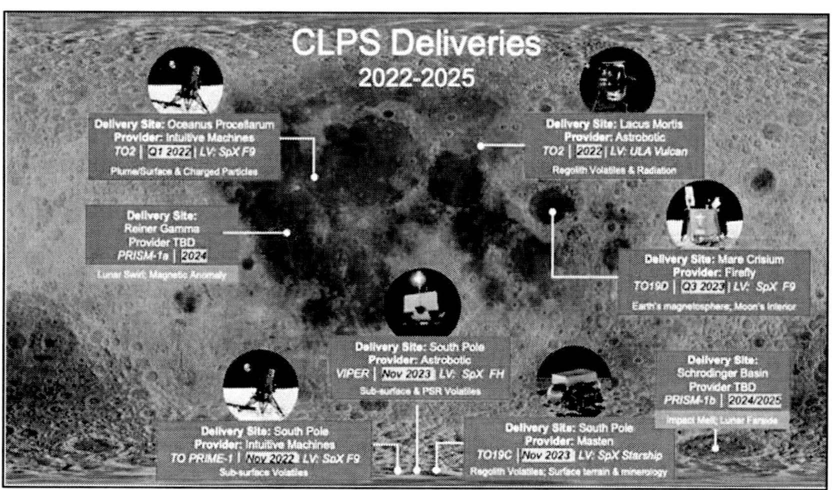

Commercial Lunar Payload Services – Destinations and Payloads.

A strong technical workforce is critical to U.S. industries, requiring a robust pipeline of talent from students to institutional academia to eventual STEM careers. The Artemis missions represent humanity's first return to the moon in nearly 50 years, and the majority of student-age and early-career STEM professionals have no direct memory of the moon landings or their far-reaching cultural and social impacts. As such, Artemis is a once-in-a-generation opportunity to recruit scientific and engineering talent into U.S. academic, government, and private industries.

The ability of U.S. firms, especially those involved in high-technology areas, to attract and retain top talent both domestically and abroad is a cornerstone of our national competitiveness.

Market conditions suggest a firm's success may be predicated on having robust revenue and investment strategy and a sophisticated and deep customer base. The influx of private equity, angel or venture capital, and SPAC-supported and other corporate investment into the private sector is currently at an apex (approaching $9 billion in 2020), with the U.S. accounting for over 64% of all investment worldwide [13]. Although the market for lunar payloads and services is not as strong now as in other space subsectors (e.g., launch, sat com), the U.S. market is more robust than any in the world. This robustness is in part due to multi-year government contracts for lunar services, which currently represent up to $5 billion, by far the most significant funding for private sector lunar missions globally. Early government support in the form of services contracts and R&D funds has historically been critical for companies to mature past the high-risk phase of development until private and corporate financing is willing to step in.

Finally, as noted above, the government's role can be critical in early stage technology development. NASA has taken a strong position supporting a broad array of competitive firms and funding service contracts to enable mission development. Artemis is mission-oriented innovation, with government proposing a strategic mission and American firms stepping up to deliver. NASA creates missions (and the enabling policy and regulatory environment) but does not create or fix markets. The "moonshot" approach to innovation, pioneered by the Apollo program, suggests that strong government support for radical technology development to tackle unsolvable problems can yield major change. As a driver for economic growth, few levers are more powerful than creating new high-technology sectors of the economy in which firms can flourish. Government missions as varied as spaceflight, weapon systems, and energy storage led to the miniaturization of hard drives, microprocessors, improved lithium battery storage, HTTP/HTML code, and capacitive screens. Adopted by Apple and smartphone makers, these innovations revolutionized communications and IT industries, generating trillions in market value [14].

One first step in establishing the legal and policy grounds for ownership of space resources will take place in the early 2020's. In August 2021, as part of an agreement between NASA and the firm Lunar Outpost to provide 4G communications as part of the Artemis program, Lunar Outpost also agreed to scoop up a sample of lunar regolith, take a picture, send it back to Earth and transfer ownership of the dust to NASA. The goal is to conduct the first sale of resources in space. For the most part, major commercial prospects for such endeavors currently remain in the well into the future. Speculating out to the 2030's and beyond, we might envision entirely new markets emerging. Once companies have established a toehold on the lunar surface and in cislunar orbit, they may provide services such as transportation to destinations in the Earth-Moon system, electrical power, rocket fuel depots, water extraction, and resource mining.

A recent surge in startups and financing support is tempered by forecasts suggesting significant commercial revenue may be beyond a 15-year window in most cases [15]. Yet other nations are pursuing expansive plans for long-term lunar exploration and research bases, bringing the moon and cislunar space into Earth's economic sphere. A U.S. national exploration strategy should be no less visionary. Future systems of governance around the moon and in Earth orbit are likely to be modeled after those nations with a commanding presence. Ideals that the U.S. champions on Earth today, including the expansion of liberal free market democracy, may help set the stage for humanity's future in space.

Conclusion

The Artemis program is likely to generate significant economic impacts, improve innovation and technology development in the space sector, and contribute to the growth of the U.S. STEM pipeline. The market environments of the COTS, CRS, and CCrew programs may be different from those encountered by the lunar commercial programs. However, the lunar programs are expected to mature over a longer time horizon. The most important demand drivers in the near term are likely to be NASA and other government agencies. Significant non-government revenues are likely to be realized from upgraded launch capabilities and demonstrated lunar crew landing capabilities developed as part of the early Artemis missions.

References

[1] https://www.nasa.gov/content/commercial-lunar-payload-services-overview; https://www.nasa.gov/nextstep/humanlander2; https://www.nasa.gov/content/nextstep-overview.

[2] https://www.ida.org/research-and-publications/publications/all/d/de/ demand-drivers-of-the-lunar-and-cislunar- economy.

[3] Falcon Heavy currently offers approximately ~$1,400-2,300 to LEO, the lowest in the world by a factor of two to three times. Source: https://www.spacex.com/ media/Capabilities&Services.pdf.

[4] https://www.nasa.gov/sites/default/files/atoms/files/option-a-source-selection-statement-final.pdf.

[5] https://inspiration4.com.

[6] (2018) Tassey, Greg. "The Economic Rationales and Impacts of Technology-Based Economic Development Policies". https://www.nist.gov/system/files/documents/ 2021/01/15/DOMESTIC%20TBED%20ProgramsPolicy%20Targets%20and%20Impacts.pdf.

[7] For example, Radiant startup founded by former SpaceX engineers: https://www.powermag.com/former-spacex-engineers-tout-new-micro reactor/.

[8] https://www.nasa.gov/directorates/spacetech/game_changing_development/projects/ SynBio.

[9] https://madeinspace.us/blog/2020/06/22/how-in-space-manufacturing-will-impact-the-global-space-economy/.

[10] https://www.nasa.gov/mission_pages/station/research/news/iss-20-years-20-breakthroughs

[11] https://www.nasa.gov/sites/default/files/atoms/files/2020_nasa_eir_brochure_for_fy1 9.pdf.

[12] Tassey, Greg (2018). https://www.nist.gov/system/files/documents/ 2021/01/15/DOMESTIC%20TBED%20ProgramsPolicy%20Targets %20and%20Impacts.pdf.

[13] https://www.cnbc.com/2021/01/25/investing-in-space-companies-hits-record-8point9-billion-in-2020-report.html.

[14] https://www.oecd.org/naec/NAEC_Mazzucato.pdf.

[15] https://www.technologyreview.com/2019/06/26/134510/asteroid-mining-bubble-burst-history/.

Index

C

D

E

F

H

I

L

M

In: Back to the Moon
Editor: Silas Blackwood
ISBN: 979-8-89530-861-5

N

P

R

S

T

V

W

Y